NASA
MOON MISSIONS
1969–1972 (Apollo 12, 14, 15, 16 and 17)

COVER IMAGES: Top: Apollo 16 Commander John Young jumps for joy and salutes the flag; Bottom, left to right: Apollo 12 crew Conrad, Gordon and Bean open a new era in Apollo after the first landing; an Apollo Command Module during assembly; the Scientific Instrument Module flown on the last three Moon missions. *(NASA)*

Dedication

This book is dedicated to NASA Deputy Administrator Robert Seamans – who, despite cynics, pushed it through and made Apollo an outstanding success.

First published in June 2019
Reprinted July 2019 and April 2023

A catalogue record for this book is available from the British Library.

ISBN 978 1 78521 210 9

Library of Congress control no. 2018953078

Published by Haynes Group Limited,
Sparkford, Yeovil,
Somerset BA22 7JJ, UK.
Tel: 01963 440635
Int. tel: +44 1963 440635
Website: www.haynes.com

Haynes North America Inc.,
2801 Townsgate Road, Suite 340
Thousand Oaks, CA 91361

Printed in India.

Acknowledgements

The author would like to thank Steve Rendle, James Robertson, Dean Rockett and Ann Page for supporting the production of this book. They are responsible for the good bits – the rest are my responsibility alone.

David Baker

Note: *The distances in this book are recorded as statute miles and not nautical miles, the latter unit being the one used in all official NASA documents. This is because most readers will be able to more readily relate to the statute mile (5,280ft) than the nautical mile (6,060ft), which is why the author has converted all distances from the nautical miles used in official reports to statute miles.*

NASA
MOON MISSIONS

1969–1972 (Apollo 12, 14, 15, 16 and 17)

Operations Manual

An insight into the engineering, technology and operation of
NASA's advanced lunar flights

David Baker

Contents

OPPOSITE The last Apollo astronaut on the Moon, Eugene Cernan, ending an age of exploration that saw humans take their first tentative steps into the Solar System. *(NASA)*

Introduction

The six Moon landings by Apollo astronauts between 1969 and 1972 were precious events in a decade of uncertainties, conflict and global tensions in the depths of a Cold War. They emerged from a technological competition to demonstrate ideological supremacy, driven by a fight for loyalty among uncommitted nations around the world. But they inspired a generation and continue to amaze the several billion people on this planet who were not alive when men first went to the Moon.

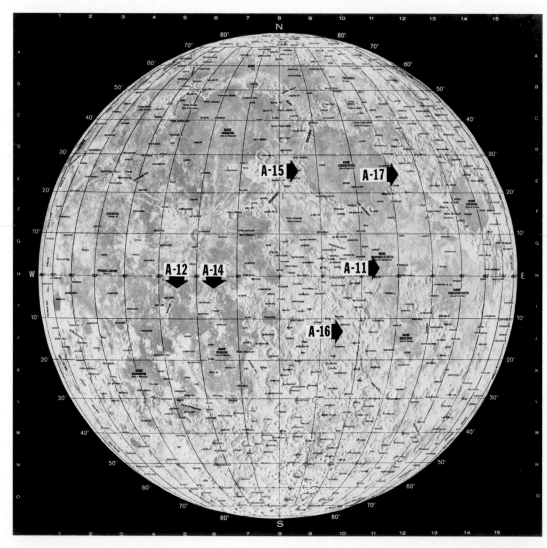

RIGHT The six Apollo lunar landing sites where 12 humans from Earth first began to explore other worlds in space. *(NASA)*

In the fifty-eight years of human space flight, more than 560 people have been into space but only 24 have left low Earth orbit and entered the gravitational attraction of the Moon. Only 12 of those have walked on the lunar surface. It may yet be many years before humans return to Earth's nearest celestial neighbour.

Many books have been written about the experiences of those who went to the Moon and about the people who helped send them on their way. Much has been written too about the science they carried out and the research they conducted. This book is different in that it explains the technology which underpinned the expansion of lunar exploration beginning with Apollo 12 in November 1969 and ended thirty-seven months later with Apollo 17. In that regard it is concerned with the engineering development of the later flights and of the modifications and improvements that made extended stays possible.

The author starts from the contentious premise that all missions up to and including Apollo 11 were qualification flights to demonstrate capability. That is just as much so for the actual first landing as it was for any of the preceding missions that led to those first steps taken by Neil Armstrong and Buzz Aldrin in July 1969. Accordingly, while the early development flights are described here in brief, their function is to provide a base from which the reader can assess the extraordinary growth in capability, and the operational flexibility which characterised the later flights.

Nowhere is that exemplified better than in the return of lunar samples to Earth, increasing with each mission, bringing back a total sample collection weighing 839.87lb (410.95kg), as follows:

Apollo 11	47.51lb (21.55kg)
Apollo 12	75.62lb (34.30kg)
Apollo 14	94.35lb (42.80kg)
Apollo 15	169.10lb (76.70kg)
Apollo 16	209.89lb (95.20kg)
Apollo 17	243.40lb (110.40kg)

Testimony too arises from the extensive set of scientific instruments left on the lunar surface operating for several years and transmitting to Earth a wealth of data that has fuelled interest in lunar and planetary exploration. Out of which a new forging of interdisciplinary sciences has emerged, unforeseen when the political objective was made to send humans to the Moon.

Several of the instruments left at the five sites where the Apollo Lunar Surface Experiments Package (ALSEP) was deployed expired prematurely but the majority continued to operate, sending back information before they were switched off on 30 September 1977.

Enshrined within the story of rapid engineering development during the Apollo years described in this book are the 24 men who flew to the Moon:

Apollo 11	Neil Armstrong
	Buzz Aldrin
	Michael Collins
Apollo 12	Pete Conrad
	Al Bean
	Dick Gordon
Apollo 14	Alan Shepard
	Ed Mitchell
	Stu Roosa
Apollo 15	Dave Scott
	Jim Irwin
	Al Worden
Apollo 16	John Young
	Charlie Duke
	Ken Mattingly
Apollo 17	Gene Cernan
	Jack Schmitt
	Ron Evans

Remembered also are the men and women of NASA, industry, science and academia who worked on the Apollo programme and helped define the achievements in this book.

David Baker
East Sussex
March 2019

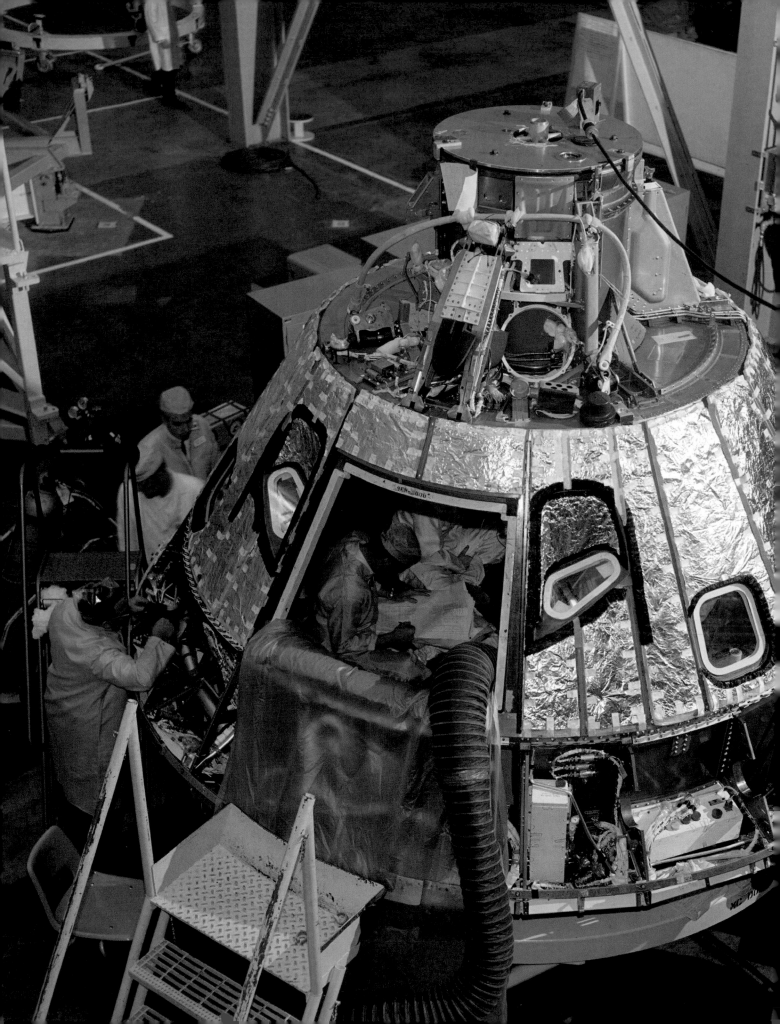

1 Prelude

When the first Moon landing by humans was accomplished by Apollo 11 in July 1969, there was sufficient hardware for a further nine missions to the lunar surface. For political and unavoidable reasons, only a further five would be flown. But the results were outstanding and far beyond anything that could have been accomplished by Apollo 11. That would take time to achieve.

OPPOSITE The design of the Apollo Command Module had all its systems placed outside the pressurised crew compartment, making it easier to manufacture and placing potentially hazardous propellants and pressurised gases outside the area where the crew would live. *(NASA)*

ABOVE President John F. Kennedy proclaimed the Moon landing goal on 25 May 1961 before a joint session of Congress. Here, on 12 September 1962, he visited NASA's Manned Spacecraft Center and viewed an early mock-up of the Lunar Module. *(NASA)*

The human space flight programme was in its infancy when President John F. Kennedy set the Moon goal as a national objective through a public declaration on 25 May 1961. Officially in being since 1 October 1958, NASA had taken control of a one-man space capsule named Mercury. Officially developed by the Air Force in the year before NASA came into existence, it had yet to put a man in orbit and only three weeks before had sent one of its first seven astronauts, Alan B. Shepard, on a suborbital ballistic shot lasting 15 minutes.

At the time, NASA had not decided how the Moon landing was to be achieved and considered four methods, or modes: Direct Ascent (DA), Earth Orbit Rendezvous (EOR), Lunar Surface Rendezvous (LSR) or Lunar Orbit Rendezvous (LOR).

The DA mode required a very powerful rocket to launch a single spacecraft direct from Earth to a direct landing on the Moon. The EOR method called for smaller launch vehicles to assemble in Earth orbit a spacecraft which could either go direct to a landing or enter lunar orbit first before descent to the surface. The LSR mode pre-positioned a spacecraft for the return trip before a lander took astronauts to the surface. Given little credence, the LOR mode had been discussed for some years and called for a single intermediate-size rocket to launch two spacecraft: one to enter lunar orbit and the other to take astronauts down to the surface.

After considerable and heated debate, the LOR mode was selected in June 1962 and by the end of that year Grumman Aircraft had been contracted to build the Lunar Module. Apollo, the spacecraft that would carry astronauts all the way to lunar orbit and back, had been studied for a year before Kennedy's decision, optimised as a successor to the one-man Mercury spacecraft and conceived to carry three astronauts for two weeks on Earth orbit science missions, circumlunar flights and, with a landing stage, to put men on the Moon sometime during the 1970s.

For the first two years NASA was confronted by a major expansion and began to slip behind the schedule required to achieve Kennedy's goal. Late in 1963 major managerial changes equipped the agency for the monumental task ahead and a large number of US Air Force officers were brought in from the missile world to transform the agency into a slick, high performance outfit. Among those brought in were George Mueller to head the manned space flight office and General Sam Phillips to manage the Apollo programme.

By early 1964 the management structure of NASA had been placed on a surer footing, the major decisions had been made regarding the mission mode and the contractors were all in place to build the Saturn launch vehicles, the Apollo spacecraft and the Lunar Module. But the LM was at least a year behind Apollo and would endure a protracted development period during which its configuration changed appreciably and the technical aspects of the engineering design went through numerous changes.

In the meantime NASA completed 12 Gemini missions, a programme introduced at the end of 1961 to develop rendezvous and docking and flights lasting up to two weeks, such as those conducting Moon missions, and space walking – Extra-Vehicular Activity (EVA). To expand the crew base, in addition to the seven Mercury astronauts selected in April 1959 NASA recruited nine more in 1962, and a further 14 in 1963. During this time, the one-man Mercury programme completed four manned orbital flights, the first on 20 February 1962 when John Glenn made four orbits of the Earth, and the last when Gordon Cooper started a 34hr flight on 15 May 1963. Then it was time to start flying the two-man Gemini, a direct development of Mercury.

The Gemini programme proved to be an outstanding success, two unmanned flights in April 1964 and January 1965 preceding the first manned mission in March 1965. Coming 22 months after the last of four Mercury missions, astronauts Grissom and Young flew NASA's first two-man spacecraft for a three-orbit shakedown flight, followed in June by a four-day flight by McDivitt and White in which the latter conducted the first US spacewalk, albeit three months after the Russians achieved the last of their significant 'firsts' before the Apollo era when Alexei Leonov went outside his spacecraft, Voskhod 2.

This was followed by eight more Gemini flights, the last of ten manned missions in November 1966 having successfully accomplished several rendezvous and docking flights, worked through troublesome problems with EVA exercises and demonstrated that humans could remain in space for two weeks – more than enough time to get to the Moon and back.

Key technologies had been demonstrated during Gemini, including the use of fuel cells for electrical power, hydrogen and oxygen fed across a nitrogen catalyst producing power with water as a by-product. Fuel cells had been selected for Apollo and this demonstration raised confidence that it was a reliable concept and could save weight compared with batteries which would have been unacceptably heavy for the power requirements of the three-man spacecraft.

Other operational workouts included the use of an improved tracking and communication network which would serve Apollo, the latter supported by improvements to the Deep Space Network (DSN) which had been built to support planetary spacecraft sent to the Moon, Venus and Mars. Tested too was the

ABOVE Initial planning for Apollo Moon missions were based around the Saturn C-1, later redesignated Saturn I, the first example of which had been launched on 27 October 1961. In the Earth Orbit Rendezvous concept several rockets of this class would have been required for a single landing on the Moon. *(NASA)*

LEFT Under the original plan, NASA would conduct a Direct Ascent to the lunar surface using an Apollo spacecraft equipped with a special landing section. This was problematical in that the forward-facing windows on the conical Command Module would have required a periscope system to provide a clear view of the surface for touchdown. *(NASA)*

newly built Manned Spacecraft Center outside Houston, Texas, with flight control teams and simulation capabilities that could refine the evolution of operations and help build a more robust support system. This was made all the more crucial after it had been decided that all the primary guidance and navigation commands would be determined through ground networks uplinked to the crew by voice for manual input to the on-board Apollo Guidance Computer.

Selection of the LOR mode sealed the fate of much bigger launch vehicles that would have been required for the Direct Ascent mode. Ex-German rocket scientist Wernher von Braun was director of NASA's Marshall Space Flight Center (MSFC) which was developing the Saturn I, Saturn IB and Saturn V rockets. The first Saturn I had flown on 27 October 1961, producing a lift-off thrust of 1.32million lb (5,871kN) but with only dummy upper stages. This S-I first stage would be developed and eventually deliver a nominal 1.6million lb (7,116kN).

The first orbital flight of a Saturn I occurred on 29 January 1964, when the fourth Saturn I carried an S-IV second stage which placed itself and a payload in orbit. This was the first flight of the high-energy cryogenic, liquid hydrogen/liquid oxygen stage powered by six RL-10 rocket motors delivering a total thrust of 90,000lb (400kN). But the Saturn I was not sufficiently powerful to put an Apollo spacecraft in orbit.

A development of the S-IV, the larger cryogenic S-IVB stage was equipped with a single J-2 engine delivering a thrust of 200,000lb (889.6kN) and this was first launched on top of a fully developed S-I first stage in a new configuration known as the Saturn IB. Its first flight was on 26 February 1966, under the designation AS-201 (all Saturn IBs were numbered in the 200-series). Capable of putting 45,000lb (20,412kg) into orbit, it would be used for early Apollo flights with the Block I and initial Block II configurations. The thrust of the

LEFT The possibility of developing a vehicle with five times the thrust of a Saturn C-5, later Saturn V, opened the plausibility of using a single rocket for each Moon landing with the Lunar Orbit Rendezvous technique. *(NASA)*

RIGHT NASA held a press conference on 11 July 1962 to announce its decision to go with the LOR mode. At extreme left is NASA Administrator James Webb. *(NASA)*

J-2 would increase over time, achieving a level of 230,000lb (1,023kN) at peak in the Apollo programme, although the engine itself would have a long and successful application on other rockets right through to the present day.

The J-2 was also employed in the Saturn V, with five in the cryogenic S-II second stage and one in the S-IVB third stage, both on top of the S-IC first stage with its five powerful F-1 engines delivering a nominal 7.5million lb (33,360kN). The first unmanned Saturn V (AS-501, all Saturn Vs being in the 500-series) was launched as Apollo 4 on 9 November 1967, followed by the second unmanned flight, Apollo 6, on 4 April 1968. Only the Saturn V could launch both Apollo CSM and LM combined.

BELOW The one-man Mercury capsule (bottom) carried NASA's first astronauts into orbit between 1961 and 1963 but the Moon decision prompted a 'stretched' version, the Gemini spacecraft. With ten manned launches in 1964–65 to develop long-duration flight, space-walking and rendezvous and docking, it was succeeded by the Apollo spacecraft which first flew a crew to orbit in 1968. *(NASA)*

ABOVE The decision to go to the Moon required construction of a massive launch complex at Cape Canaveral on leased land designated as NASA's Kennedy Space Center. That included the giant Vehicle Assembly Building capable of processing four Saturn V rockets simultaneously. *(NASA)*

Building Apollo

The full story of the technical development of the Apollo spacecraft is far too extended and complex to do more than summarise here but there are some features of the spacecraft that bear witness to its chequered history insofar as its mission objectives are concerned. These are important to note as readers will recall that the spacecraft was conceived early in 1960 as a successor to Mercury and that it was designed in conceptual form for a somewhat different purpose. Following what had been known as the 'von Braun' paradigm of space station, circumlunar, lunar landing and planetary exploration in that order, originally Apollo was intended to provide a multi-purpose role both for extended Earth-orbit science missions and for circumlunar flights.

From the outset the basic design incorporated a pressurised Command Module (CM) in which the crew would spend their time working, resting and eating, and a cylindrical Service Module (SM) which was not pressurised but housed all the support systems required for extended orbital or circumlunar flight. It would be jettisoned prior to re-entry and burn up in the atmosphere. Decisions as to which systems to choose, and the specification for each, were driven by the mission requirements which were in turn subject to the changes brought about by the Kennedy decision.

One of those installed systems that was never changed when the LOR decision was made, removing the Apollo spacecraft from touchdown duty, was the Service Propulsion System, the main rocket motor in the SM which was required to have a thrust of 20,000lb (88.96kN). Under 1/6th lunar gravity, this would be capable of lifting a 1g mass of 35–40 tonnes off the surface of the Moon. This excessive velocity capacity was to prove useful in the extended missions of the later Apollo landings.

A crucial decision was to use fuel cells in the Apollo CSM but to use batteries in the Lunar Module, despite early requirements for a fuel cell electrical production system in the lander too. Weight and other considerations induced a change from that to a more conventional electrical supply, made possible due to the operating life of the LM being 72 hours versus 14 days for the Apollo spacecraft. But the consideration of the use of the LM as a lifeboat in the event of a failure in the CSM up to the point of descent for landing influenced its specification and design to some considerable extent.

The possibility of an LM stranded on the Moon was an enduring nightmare at NASA and had stimulated various proposals for equipment that might be used as an emergency lift-off to lunar orbit, none of which proved practical. However, the use of the LM Descent Propulsion System (DPS) – the main engine used to descend to the surface – as a stand-in for a failed SPS influenced the size of that motor's propellant tanks, which were extended in volume by length to provide the necessary delta-velocity

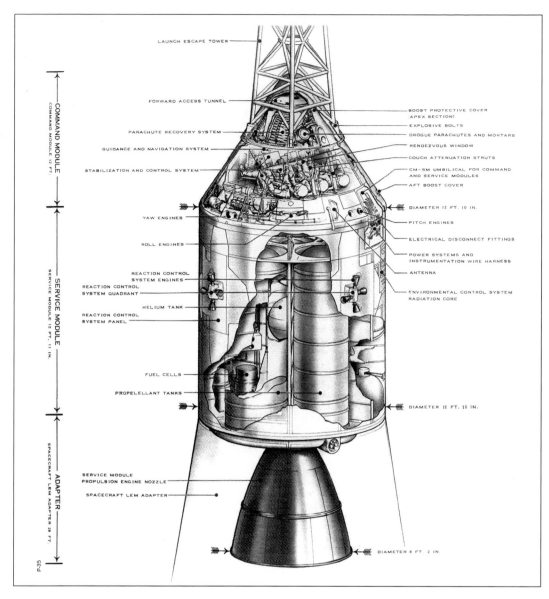

LAUNCH ESCAPE TOWER

FORWARD ACCESS TUNNEL

PARACHUTE RECOVERY SYSTEM

GUIDANCE AND NAVIGATION SYSTEM

STABILIZATION AND CONTROL SYSTEM

YAW ENGINES

ROLL ENGINES

REACTION CONTROL SYSTEM ENGINES

REACTION CONTROL SYSTEM QUADRANT

HELIUM TANK

REACTION CONTROL SYSTEM PANEL

FUEL CELLS

PROPELELLANT TANKS

SERVICE MODULE PROPULSION ENGINE NOZZLE

SPACECRAFT LEM ADAPTER

BOOST PROTECTIVE COVER (APEX SECTION)

EXPLOSIVE BOLTS

DROGUE PARACHUTES AND MORTARS

RENDEZVOUS WINDOW

COUCH ATTENUATION STRUTS

CM–SM UMBILICAL FOR COMMAND AND SERVICE MODULES

AFT BOOST COVER

DIAMETER 12 FT. 10 IN.

PITCH ENGINES

ELECTRICAL DISCONNECT FITTINGS

POWER SYSTEMS AND INSTRUMENTATION WIRE HARNESS

ANTENNA

ENVIRONMENTAL CONTROL SYSTEM RADIATION CORE

DIAMETER 12 FT. 10 IN.

DIAMETER 8 FT. 2 IN.

COMMAND MODULE

COMMAND MODULE 12 FT.

SERVICE MODULE

SERVICE MODULE 12 FT. 11 IN.

ADAPTER

SPACECRAFT LEM ADAPTER 28 FT.

P-35

for such an eventuality. That proved useful
during the failed Apollo 13 mission, giving the
stricken mission additional velocity potential.

Other changes to the Lunar Module design
came as a result of an evolution in both
thinking and technical refinement and have
been told elsewhere in numerous places in
books and on the Internet. Suffice here to
say that the enthusiasm for extended Apollo
surface operations peaked during major
engineering development and influenced the
need for stowage locations which were put
to good use for the Apollo Lunar Surface
Experiments Package (ALSEP), the set of
experiments left on the lunar surface, and the
Lunar Roving Vehicle (LRV).

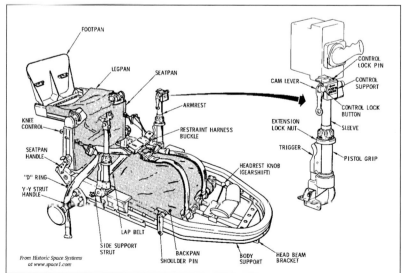

FOOTPAN

LEGPAN

SEATPAN

ARMREST

KNEE CONTROL

RESTRAINT HARNESS BUCKLE

SEATPAN HANDLE

"D" RING

Y-Y STRUT HANDLE

SIDE SUPPORT STRUT

BACKPAN SHOULDER PIN

LAP BELT

HEADREST KNOB (GEARSHIFT)

BODY SUPPORT

HEAD BEAM BRACKET

CAM LEVER

CONTROL LOCK PIN

CONTROL SUPPORT

CONTROL LOCK BUTTON

EXTENSION LOCK NUT

SLEEVE

TRIGGER

PISTOL GRIP

From Historic Space Systems at www.space1.com

At the time the LM was designed neither of these payloads were planned. Along the way, simplifications were introduced, including adopting two propellant tanks in the Ascent Stage, one for fuel and one for oxidiser, rather than the four originally planned. Because the two propellants have very different mass/volume values, this necessitated the tank on one side being cantilevered further out, an offset installation necessary to balance the different weights and align the centre of mass.

The problem of how to employ the Apollo CSM as a Moon lander had been problematical but essentially the basic specification worked up prior to the Kennedy decision set the initial design requirement, with a descent propulsion system incorporated into a landing leg structure. The decision to change from Apollo as a lunar lander to a lunar-orbiting mothership came fully a year after North American Aviation had begun work on the contract. The sequential development of a Block A spacecraft, for early manned flights in Earth orbit supporting initial verification of operational systems and qualification of essential design features, was renamed Block I. These flights were to be followed by the Apollo B configuration, becoming the Block II in January 1964 with full deep-space specification including a docking capability. But in reality the differences were profound.

Block I evolved into a production spacecraft before the decision to adopt the LOR mission mode and had an arrangement of equipment and systems resulting from its initial purpose. Because it was never intended for such operations in the Direct Ascent mode, Block I was not configured for rendezvous and docking and had only a small forward access tunnel, with no docking equipment, which would provide a second means of egress should the side hatch become inoperable after splashdown. The side hatch itself was a two-piece structure with an internal hatch bolted to the interior and an external hatch which took some time to unlock after the internal hatch had been removed and placed inside the spacecraft.

The Command Module was encapsulated on the pad and during the early phase of ascent by a Boost Protective Cover (BPC), a cone that covered the exterior of the Command Module to the point where it mated to the Service Module which incorporated what constituted a third hatch which also had to be removed for ingress or egress. All this was simplified into a single integrated hatch for Block II while retaining the BPC.

Block I had a much smaller umbilical connector between the Command and Service Modules and was located close to the side hatch whereas the connector for Block II was much larger and located on the opposite side to the ingress/egress hatch. Several changes were made to the arrangement of the VHF antennas and the Block II spacecraft also had a unified S-Band (USB) antenna located on an arm at the base of the Service Module for deep-space communication.

The Block II Service Module had significant differences in the location of the radiators for the electrical production system (EPS) and for the environmental control system (ECS) as well as in the positioning of the cryogenic reactant tanks and the three fuel cells. There were changes too in the shape and configuration of the thermal shielding at the base end of the SM, effectively to protect it from heat generated by the expansion nozzle for the SPS engine.

These Block differences were minimal visually and only an eye accustomed to looking at Apollo spacecraft would notice the external disparities but the colour was a

BELOW Guidance, Navigation & Control System (GN&CS) equipment was placed on the sidewall of the Command Module with sextant and telescope penetrating the pressure vessel and the exterior heat shield. *(NASA)*

EYEPIECE STOWAGE
SIGNAL CONDITIONER ASSEMBLY
OPTICS ASSEMBLY
SCANNING TELESCOPE
SPACE SEXTANT
INERTIAL MEASUREMENT UNIT
CONTROL PANELS
POWER SERVO ASSEMBLY
COUPLING DATA UNIT
COMMAND MODULE COMPUTER
NAVIGATION BASE

SIDE VIEW

significant giveaway: the Block I spacecraft flown on Apollo 6 had a white painted finish on the SM and the CM but on all other flights the SM was a natural metal finish with the radiators painted white.

At the core of the Apollo programme was a concern over quality control and safety, with more than one 'whistle-blower' alerting programme managers to issues at both the contractors and at some NASA field centres where the pressure to deliver was remorseless.

One of these issues concerned the issue of pure oxygen in the spacecraft, which had been selected on the basis of a trade-off regarding the overall reliability (and safety) of a two-gas versus a one-gas system. The technical arguments for rejecting an oxygen/nitrogen environment were complex but essentially they stated that the use of a mixed-gas atmosphere would introduce a level of risk from multiple subsystems, increase weight and over-engineer the system. As a result, a pure oxygen environment was decided upon.

General Electric had been under contract to carry out quality assurance programmes and GE's Hilliard Page had sent Joseph Shea, deputy director of manned space flight, a letter dated 29 September 1966, alerting him to dangers his team believed existed over the potential consequences of a fire in such an oxygen-rich atmosphere. Shea sent the letter to William Bland at the Reliability, Test & Quality Control Division of the Apollo programme office who was overwhelmed with work and failed to respond immediately to that concern, not returning a response for two months, following which Shea wrote to Paige on 5 December asserting there was an 'adequate margin of safety' but that 'the problem is sticky – we think we have enough margin to keep the fire from starting – if one ever does, we do have problems. Suitable extinguishing agents are not yet developed'.

At the root of the concern was the measured increase in flame propagation rates across materials soaked in a pure oxygen atmosphere and made exponentially worse at increased pressure, bad enough at the 5lb/in² (34.47kPa) of the cabin in orbit but terrifyingly combustive at sea-level pressure of 14.7lb/in² (101.3kPa). There were numerous examples of these effects

LEFT This view of a Block I Apollo spacecraft shows the relative location of the Command Module and its Service Module, the latter containing all the environmental control and life support systems, together with the electrical production system, the Service Propulsion System (SPS) engine, the Reaction Control System (RCS) thrusters, communications equipment in addition to passive and active thermal control. (NASA)

under test and controlled experiments but in defence of the status quo NASA had flown 16 manned space flights without a single issue

BELOW Communications between Earth and Apollo was a vital part of up-linking data and maintaining contact with the crew via an S-Band antenna located on the aft surface of the Service Module. Supported by an articulating arm, the four-dish antenna could rotate and align with Earth. (NASA)

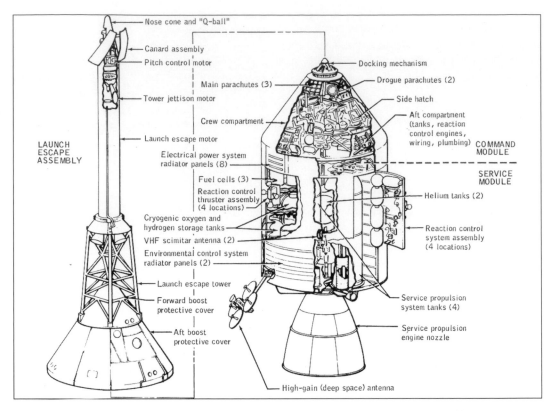

RIGHT The Block II Apollo CSM as configured for all manned flights after the Block I was downgraded for uncrewed missions only. *(NAA)*

Nose cone and "Q-ball"

Canard assembly

Pitch control motor

Docking mechanism

Main parachutes (3)

Drogue parachutes (2)

Tower jettison motor

Side hatch

Crew compartment

Aft compartment (tanks, reaction control engines, wiring, plumbing)

LAUNCH ESCAPE ASSEMBLY

Launch escape motor

COMMAND MODULE

Electrical power system radiator panels (8)

SERVICE MODULE

Fuel cells (3)

Helium tanks (2)

Reaction control thruster assembly (4 locations)

Cryogenic oxygen and hydrogen storage tanks

Reaction control system assembly (4 locations)

VHF scimitar antenna (2)

Environmental control system radiator panels (2)

Launch escape tower

Forward boost protective cover

Service propulsion system tanks (4)

Aft boost protective cover

Service propulsion engine nozzle

High-gain (deep space) antenna

RIGHT The Block II Apollo CSM as configured for all manned flights after the Block I was downgraded for uncrewed missions only. *(NAA)*

BELOW The Apollo Service Module contained the electrical power production system, comprising three fuel cells operating on a reaction of hydrogen and oxygen over a nitrogen catalyst in a process of reverse electrolysis. The cryogenic reactants were contained in four tanks, the hydrogen tanks being sheathed in a circular jacket. *(NASA)*

Fuel cell 2

Fuel cell 3

Fuel cell 1

Fuel cell shelf

Oxygen tank 1

Oxygen tank 2

Oxygen valve module

Oxygen subsystem shelf module

Oxygen fuel cell shut-off valve module

Oxygen servicing panel

Hydrogen tank 1

Hydrogen subsystem shelf module

Hydrogen tank 2

and believed it could manage pure oxygen in Apollo. Paige was not satisfied and considered reporting it higher up over Shea's head but decided not to.

There were other issues too, with much the same over quality control and shoddy workmanship, and issues raised by an employee of North American Aviation (NAA), working the Apollo CSM and Saturn V S-II stage contracts. Concerned at what he believed were serious flaws, Thomas Baron was employed by North American but had repeatedly made an issue over what he believed were seriously flawed work practices. When it came to alerting NASA over the heads of his bosses in late 1966, the issue got some traction but then he was sacked and had to meet with NASA elsewhere. This raised serious concerns but there was little that could be done at that stage to turn the entire culture around. But it came a year after a much more robust concern from the very top of the command chain.

Late in 1965, Mueller asked Phillips to investigate concerns he had regarding the late delivery of certain items, noting the schedule slipping across the board and serious cost overruns. NAA was working both the CSM and S-II stage contracts for the Saturn V and the technical troubles with the latter were to some measure understandable but the repeated changes made by NASA also compromised the satisfactory completion of Apollo spacecraft on what was a hand-built, numerically limited production line. Nevertheless, with concerns expressed across a wide front there were serious misgivings about the two contracts and Phillips sought answers from the President of NAA, Leland Atwood.

Phillips organised a group on 22 November to visit, make reports and formalise exchanges on a series of intensive inspections through 6 December. As a result of notes taken during those visits, Phillips sent a letter to Atwood dated 19 December 1965, saying in part that 'even with due consideration of hopeful signs, I could not find a substantive basis for confidence in future performance. The gravity of the situation compels me to ask that you let me know, by the end of January if possible, the actions you propose to take'.

However, with the end of Gemini flight operations in November 1966 it was time to fly Apollo, the first flight having been named Apollo 1 (AS-204), crewed by veteran astronauts Grissom and White and the rookie Roger B. Chaffee, who had been in training for their mission since selection in January 1966, except for Chaffee who had replaced Eisele when the latter dislocated his shoulder. There had been some hope of flying the first manned Apollo mission on a dual flight with the last Gemini, Apollo 1 having been scheduled for October on the delayed schedule modified as a result

of earlier slippages, although without a means of docking there was little purpose in that. But the delays to delivery of the spacecraft and the technical problems uncovered after the spacecraft arrived at the Kennedy Space Center removed any possibility of that occurring.

On 26 October 1966, consultants Bellcomm issued a report regarding the possibility of the first manned Apollo flight conducting a rendezvous with the Agena flown for the last manned Gemini mission, GT-XII. The primary reason was to retrieve experiment packages left on the exterior of the stage by a spacewalk from Apollo during a close rendezvous with the inert Agena. There was also the possibility of rendezvous by the second Apollo mission (AS-205) which at this time was anticipated as a Saturn IB launch, or on the following paired flight with an unmanned Lunar Module launched on a Saturn IB in the so-called AS-207/208 dual mission, or on the first manned flight of the Saturn V, at this date anticipated for the third flight of that rocket (AS-503).

With delays piling up, a detailed examination of the existing flight schedule brought changes to the crew assignments and to the mission objectives. The first flight was to be followed by AS-205 (Apollo 2) flown by Schirra, Eisele and Cunningham, with McDivitt, Scott

LEFT The three fuel cells supported by the upper of three separate shelves in the segmented bay of the Service Module. *(NASA)*

and Schweickart as back-up, a selection announced on 29 September 1966. Essentially a repeat of the AS-204 flight, AS-205 was to be followed by the first unmanned flight of the Lunar Module on AS-206. The dual AS-207/208 mission would be launched on separate Saturn IB flights to check out a docked operation between the separately flown CSM and LM. These five Saturn IB missions would pave the way for a manned launch of both the CSM and LM on the third Saturn V flight (AS-503) following two test flights.

Following a critique of his function as commander of the second Apollo mission, Schirra lobbied for the AS-205 flight, essentially a repeat of the Apollo 1 mission, to be removed as unnecessary, which in December 1966 it was. This resulted in Schirra's crew being reassigned as back-up to Apollo 1 and the original back-up crew for that flight (McDivitt, Scott and Schweickart) being given the dual mission aboard a Block II CSM originally designated AS-207/208 but now referred to as AS-205/AS-208, since the now vacant AS-205 launcher slot would be used for that first in-space test of both CSM and LM, albeit launched on separate Saturn IBs. Borman, Collins and Anders were assigned to fly the first

Saturn V manned launch (AS-503) as a highly elliptical Earth orbit mission with both CSM and LM. When he entered hospital for bone surgery in 1968 Collins was replaced by Lovell in Borman's crew.

All of that became academic when, on 27 January 1967, a fire broke out in the Command Module and after brief outbursts of alarm, the only truly decipherable phrase bringing the chilling words 'Fire in the spacecraft!' from a crewmember. Within a minute Grissom, White and Chaffee had asphyxiated on the gases caused by the rapid combustion of flammable materials creating a toxic atmosphere.

It was a fire made worse by the overpressure in the spacecraft – the pressure had been raised to 16.7lb/in^2 (115kPa) after the hatches had been sealed and the air replaced with pure oxygen to prevent mixed-gas air leaking in from outside. In a pad test on Saturn IB AS-204, the crew were simulating a countdown such as they would conduct for a planned lift-off on 21 February, the test scheduled to end with the practice of an emergency evacuation. But the simulation was delayed long after its scheduled duration by sustained problems and difficulties getting an effective voice signal between the spacecraft and the ground.

Spacecraft 012 had arrived at Cape Canaveral on 26 August 1966, with 113 declared planned engineering changes still to be applied. NASA personnel had to apply an additional 623 engineering changes before flight, delaying the mission by several months. Despite the problems outlined by numerous sources in the previous 18 months, the spacecraft had been delivered in an undesirable condition and the crew became frustrated over the technical hitches that continued to confound their training sessions with the flight hardware. So much so that Grissom hung a lemon on the door of the simulator.

During testing prior to the flight the environmental control system was found to be defective and had to be returned to the manufacturer, but upon return it leaked water-glycol solution and repairs were necessary. The fire risk under high pressure in a pure oxygen environment had been noted and accepted by engineers and a procedure had been put in place to mitigate that. During a two-day test beginning on 18 October 1966 the Apollo 1 crew had participated in a test whereby, for operations on the pad and during ascent through the atmosphere, pressure would be reduced to 5lb/in² (34kPa) to reduce that risk. The procedure had also been conducted on 30 December by the back-up crew of Schirra, Eisele and Cunningham. But it had not been implemented for the test on 27 January 1967.

The White House and Congress afforded NASA credibility and openness in handling the investigation itself, remarkable under the circumstances, with both Houses of Congress holding their separate hearings along with the space agency's internal investigation and analyses. The political impact was made worse by the record of concern that unravelled during the interviews with NASA and contractor executives. Embarrassed to learn for the first time about the report by Sam Phillips into the performance of North American Aviation, the investigation began to disclose a record of concerns which had not been publicly disclosed; moreover, NASA Administrator Jim Webb had also been kept in the dark.

NASA's investigation resulted in a 3,000-page report which was delivered to Webb on 9 April, a day before the Congressional hearings began. Chastised by Congressional interrogators, the NASA leadership was seen to have been less than frank with its paymasters and were charged with incompetence bordering on malfeasance, in that it contravened the lawful use of public money in supporting contractors who had produced a spacecraft which was judged unsafe and unable to fulfil its objectives.

Some in Congress wanted a five-year halt in the space programme for a thorough transformation under scrutiny and oversight; many Democrats were concerned that this event would undermine NASA's ability to fulfil the directive of Kennedy and a few wanted NASA abolished and the space programme handed over to the Air Force. It was serious for the White House because this was a government department, responsible to the President for direction, run by a man who had been selected by the White House and the problems were seen as a blight on that record.

Webb argued for a two-week moratorium on Congressional action, so that he could assimilate the report more fully into planning and strategies which he would bring back to Congress. During that time, Webb met frequently with the chairs of the two committees and impressed them with his candour, personal acceptance of responsibility and with plans for putting Apollo back on track, delivering the lunar landing and restoring pride in America's space programme. It helped too that the intelligence community was briefing Congress on recent information regarding a super-heavy launch vehicle being developed by the Russians, in a year when many expected them to celebrate the Bolshevik revolution with a Moon landing attempt.

It also helped NASA having Frank Borman, an astronaut seen as open and honest, one whom politicians could trust, who went to the committees and testified using language usually reserved for diplomats: 'We are trying to tell you that we are confident in our management and in ourselves. I think the question is really,

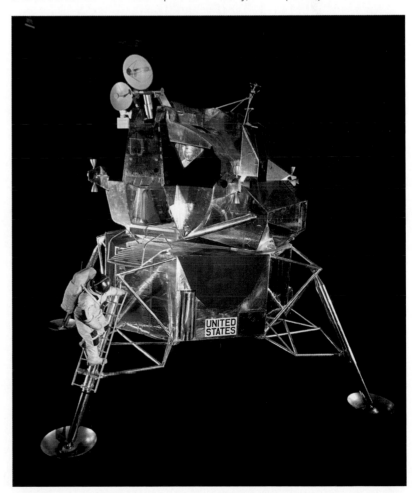

BELOW Represented here by a crew training mock-up, the Lunar Module consisted of an Ascent Stage and a Descent Stage with the pressurised crew compartment and overall structure protected by thermal insulation which gave the spacecraft a frail and fragile appearance. Under that, the structure was designed to be as light as possible, the strength of the landing legs unable to support its weight on Earth. (NASM)

RIGHT Seen here on Lunar Test Article-1 at the National Air and Space Museum in Washington DC, the asymmetric appearance of the Ascent Stage was caused by the different weights of fuel and oxidiser requiring the heavier oxidiser tank to be closer to the centre of gravity to balance the lighter fuel. The right side of the LM shown here displays the nitrogen tetroxide oxidiser tank. *(NASM)*

RIGHT Another view of LTA-1 showing the left side where the hydrazine fuel tank is cantilevered out from the pressurised crew compartment. *(NASM)*

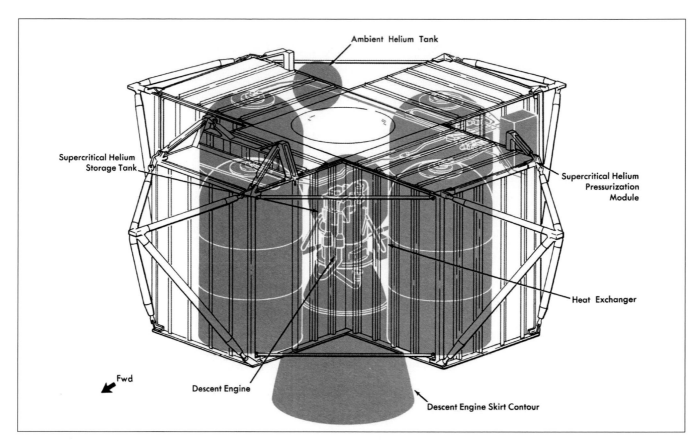

Ambient Helium Tank

Supercritical Helium
Storage Tank

Supercritical Helium
Pressurization
Module

Heat Exchanger

Fwd

Descent Engine

Descent Engine Skirt Contour

are you confident in us?' In the view of many who were present at those hearings, Borman saved the day by throwing the question back at the politicians, in effect asking them whether they had the right to change the pace and orientation of a programme over the heads of those who were right at the forefront of risk and technical experience.

The fire shattered the nerves of Joe Shea, a man already overworked and overstressed who was running on borrowed hours and little sleep. It got to him and never left him for the rest of his life. He turned to alcohol and barbiturates and was screened by a psychiatrist who judged him sane but deeply affected both by the intensity of the last several years and by the fire. NASA removed him to a headquarters job but he left within two months. At North American, Lee Atwood refused to fire Harrison Storms, head of the space division, but Webb was clear: Storms goes or the contract is revoked. He was fired.

Technical changes to the Block II spacecraft were significant and could have been made before the fire itself but several modifications were already in place. Apart from changes to the hatch, the interior was re-examined and

reappointed with materials which would not burn and could self-extinguish. Operating procedures were improved in that when the crew were on the pad, or otherwise in the spacecraft under ambient atmospheric pressure, the Command Module would be pressurised with a 60/40 oxygen/nitrogen gas, bleeding away during ascent so that when the spacecraft reached orbit, and the astronauts removed their helmets, there would be a partial pressure of 3lb/in² (20.7kPa) which is about the same as that percentage at ground level where the oxygen content is 20.9%.

Technical examination of the Command Module after the fire failed to find a single cause for the accident but the general condition of the workmanship left many potential sources for a spark or other form of ignition which rapidly spread fire across nylon webbing and netting used liberally for securing various items around the cabin. Moreover, flammable pipes and wire insulation was replaced and aluminium used instead. These changes were implemented for Block II and there were to be no manned flights with Block I, although that series would still be used for the Saturn V test flights.

ABOVE The Descent Stage provided a stable platform for landing and contained the Descent Propulsion System (DPS) which would lower the LM to the Moon. Shown here without landing legs are the two fuel and two oxidiser tanks together with the helium pressurisation system. *(Grumman)*

Recovery

Under the original naming structure, the first manned Apollo would be forever remembered as Apollo 1, in accordance with the wishes of the deceased astronauts' families, but from 24 April 1967, Mueller decreed that subsequent flights would numerically pick up with Apollo 4, scheduled as the first flight of the Saturn V (AS-501), followed by Apollo 5, the first flight of the Lunar Module (LM-1), and Apollo 6, the second unmanned Saturn V mission (AS-502). Although never officially designated as such, the two Apollo CSMs which had flown on Saturn IB missions AS-201 and AS-202, would generally be referred to as Apollos 2 and 3, respectively. Apollo 5 would use the Saturn IB (AS-204) which would have carried Apollo 1 into orbit; it was removed from Launch Complex 34 and relocated to LC-37 from where it was launched.

The delay to the Apollo flight schedule due to the fire and the application of corrective procedures caused a complete review of the sequential steps necessary to get to the first Moon landing. It needed a clearly defined set of incremental stages to flight-qualify spacecraft, systems and procedures and that was defined by Owen Maynard. A Canadian by birth and a 'refugee' from the Avro Canada CF-105 Arrow project, cancelled in 1959, Maynard was recruited by NASA to serve the Mercury programme and made a valued contribution.

In 1960 he was moved to a small group working with Bob Piland on what would become the Apollo spacecraft but in 1963 he was appointed head of the Lunar Module engineering office at the Manned Spacecraft Center. Maynard is generally credited with being more responsible than any other individual for the basic design and configuration of the LM until detailed design and development was taken up by Grumman at the end of 1962. By 1964 he was chief of the Engineering Systems Division and spent a lot of time integrating the Apollo spacecraft and working up the various development steps necessary to verify space-worthiness.

From this work, Maynard devised an alphabet-step process, assigning a separate letter for each stage necessary to build toward the lunar landing. To this date there had been an ad hoc approach, moving (as related earlier) through a logical but progressive sequence until the landing could be accomplished. But Maynard wanted something more tangible and his letter-step chart became the accepted sequence, as follows.

A Development flights with the Saturn V to qualify the launch vehicle and Apollo CSM involving deep-space elliptical trajectories for testing the Block II heat shield;

B An unmanned qualification of the Lunar Module on a Saturn IB with ground-commanded tests of the systems including propulsion;

C Manned flight of the Apollo CSM for a 10–14 day mission to thoroughly check out and verify systems and propulsion;

D Manned flight of the CSM and LM on a Saturn V for flight testing of the LM during a manned separation and rendezvous and docking exercise in Earth orbit;

E Manned CSM and LM on a Saturn V for a simulated lunar mission in an elliptical orbit with an apogee of 4,040ml (6.500km);

F Manned CSM and LM on a Saturn V for a full dress rehearsal in lunar orbit to verify around the Moon the separation, rendezvous and docking techniques;

G The first manned Moon landing.

BELOW NASA officials give testimony to Congressional hearings into the Apollo fire. From left, Deputy Administrator Robert Seamans, Administrator James Webb, manned space flight boss George Mueller and Apollo programme manager Sam Phillips. *(NASA)*

Maynard's A–G sequence was adopted as a seven-step roadmap on 20 September 1967. Based on the projected availability of the spacecraft hardware and the engineering challenges posed by flight qualification of a new spacecraft, it was possible that each step would require more than one attempt to complete its objective and the alphabetically arranged stages were milestones contingent on each one fulfilling its goals. NASA management was fully aware that it might take more than the scheduled number of flights to achieve that.

But there were concerns. The processing capacity of the Kennedy Space Center allowed for a single Saturn V launch every two months. It was already expected that there would need to be two flights of step A (to qualify the Saturn V) before moving to subsequent steps. There were at least six Saturn V-related launches involved in the five steps involving that rocket and any one of steps D to G could require one or more flights. Since the fire, when many people, several within NASA, feared that the goal was unachievable by the end of the decade, managers recognised that to get the programme back on track it was necessary to get at least the first four steps completed by the end of 1968.

But there were serious developmental problems with the S-II second stage of the Saturn V. That story has been told effectively in another Haynes Workshop Manual on the Saturn V but suffice to state here that the delay was not surprising, given that the scale of the requirement was unprecedented; the design and test of such a large liquid oxygen/liquid hydrogen rocket stage being one-of-a-kind in the 1960s. Overlaying the anticipated technical challenges were concerns regarding the performance of North American Aviation but that has been stated. Development of the cryogenic J-2 rocket motors proceeded more smoothly but the overall challenges did put back the first flight of the Saturn V as step A on Maynard's ladder.

When it came on 9 November 1967, the flight of Apollo 4 with AS-501 and Block I CSM-017 was a monumental triumph, total vindication of Mueller's all-up systems test philosophy and a great relief to the entire NASA team. The S-IVB third stage performed well, with re-ignition to propel the spacecraft to a

ABOVE Seen here in white shirt, facing the camera and holding papers, Owen Maynard established the sequential steps, each identified by a letter in the alphabet, by which NASA could achieve its Moon goal by the end of the 1960s. *(NASA)*

highly elliptical orbit qualified as a capability by the Saturn IB AS-203 flight in 1966, which had verified the operation of that stage. On returning through the atmosphere to a safe recovery, the Command Module demonstrated no serious anomalies, clearing the design of the spacecraft for deep-space operations.

LEFT Seen here delivering a briefing on the Apollo LOR decision, Joseph Shea became a victim of the Apollo fire, both career wise and psychologically. *(NASA)*

ABOVE The flight crews originally selected for early Apollo flights. Front, from left: White, Grissom and Chaffee; back from left: Scott, McDivitt and Schweickart. The fire would create many crew shuffles as the sequence of flight operations was transformed. *(NASA)*

The next development flight occurred on 22 January 1968, when Saturn IB AS-204 took the unmanned Lunar Module-1 to Earth orbit for a shakedown demonstration and test of vital spacecraft systems, most important of which were the propulsion units for the Descent Stages and the Ascent Stage. The launch vehicle performed well but there were some problems with LM-1 and serious concerns were expressed about flying only one Lunar Module on a B-series flight. Post-flight analysis showed that none of the problems encountered need impede progression direct to a manned test of the Lunar Module on a D mission. But that would not involve the second production spacecraft because LM-2 had been built in a configuration that would not support manned operations; it ended up in the National Air and Space Museum, Washington DC, where it can now be viewed as a fully flight-rated spacecraft.

Further success came with the second A flight when the second Saturn V (AS-502) took Apollo 6 into space on 4 April 1968 on what was supposed to be a repeat flight of the Apollo 4 mission but with a different set of test objectives for launch vehicle

development. That did not go at all well, when serious vertical oscillations in the second stage caused an igniter line to rupture in one of the five J-2 engines. But, an error in wiring caused the shutdown command to trigger a sequence of events which shut down a second engine as well, requiring the stage to burn much longer than anticipated as fuel was consumed by the remaining three rocket motors. After reaching orbit the S-IVB failed to restart, leaving CSM-020 to propel itself to high apogee before returning the Command Module to a safe splashdown.

In the intervening period between the fire and the first flight of a manned Apollo spacecraft, three development flights had taken place involving the newly minted Saturn V and the freshly flown Lunar Module. On two of these flights there were serious engineering problems which may have caused NASA to approach the initial manned flight with some caution. But the realistic evaluation of the anomalies did not affect any of the hardware designs being prepared for Apollo 7 – the first manned flight – and it was prudent and cautiously responsible to press ahead. In fact, the added performance call on the SPS engine in CSM-20 after the S-IVB failure gave reassurance that this piece of hardware was in every respect fit for supporting crewed flight.

But there were some problems with the Lunar Module and it became apparent that LM-3 would not be ready for the D mission involving both CSM and LM on the first manned Saturn V in what was then referred to as Apollo 8. As a result of the success with steps A and B, in August 1968, just two months prior to the first flight of a manned CSM, and recognising that LM-3 would not be ready before at least February 1969, it was decided to revise Maynard's alphabet-steps and insert a completely new mission, known as C' (C-prime) as follows:

C: The first manned Apollo CSM flight on Saturn IB AS-206 scheduled for October, a ten-day shakedown flight with Schirra, Eisele and Cunningham, formerly the back-up crew to Apollo 1;

C': A manned Saturn V (AS-503) flight to lunar orbit and back with Borman, Lovell and Anders to verify cislunar operating procedures, deep-space communications, navigation and the heat shield.

Mission D would be flown as originally planned but mission E would be deleted as unnecessary, its objectives having been achieved with C', a rearrangement that would flip two flights and rename them in a numerical missions sequence determined by date and not by objective according to Maynard's alphabet-steps. But in the year since formulating the original set of objectives a realisation emerged that perhaps, despite the delays, there would be potentially ten flights to the surface of the Moon. Already, NASA had been working through several possible developments for what it called the Apollo Applications Program (AAP).

Unlike the politicians and the public, NASA had never thought of the Apollo programme as a one-shot-to-glory flight – a bootprints and flags job for national and international acclaim. NASA plans envisaged the use of Apollo hardware to build on early landings for a continuing programme of scientific exploration, expeditions to follow in the tradition of Lewis and Clarke, James Cook and Christopher Columbus. Scientists recognised they would always play a back role in getting to the Moon but, once there, they were keen to

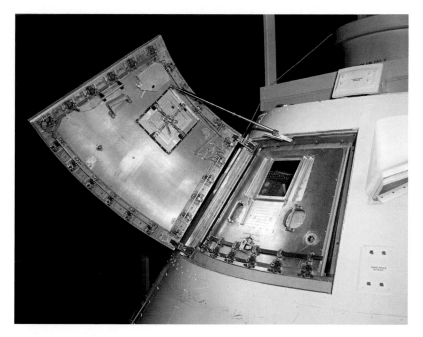

build upon early flights and establish semi-permanent bases using adaptations of the Apollo spacecraft and the Lunar Module for transporting habitats, roving vehicles and a new set of tools and equipment for expanding the geological survey of the surface.

The peak years for optimism were 1963–65 on the back of an escalating NASA budget which soared from $700million in 1961 to $5.2billion in 1965 before starting down, falling

ABOVE The triple-hatch design for access in and out of the Block I Command Module used a clumsy arrangement which took several minutes to open and this was replaced with a quick-opening hatch for all manned flights. *(NASA)*

LEFT The hatch for Apollo Block II spacecraft was operated by a single handle operating a ratchet to release, or lock, latches around the exterior. *(David Baker)*

to $3.3billion in the year NASA stopped Moon flights. Industry built 90% of the hardware operated by NASA and jobs soared to a peak of 450,000, of which 280,000 worked on the Apollo programme alone. There was justification anticipating an American space-faring future and NASA was eager to start work on those plans.

But, by using existing hardware to establish a more permanent presence on the Moon, it would have to buy more Saturn V rockets. NASA had contracted for 15 sets of launchers and spacecraft in the expectation that in the uncertainty of how many might be required

to get to the first landing, additional orders could follow as required. But the cost of those massive Saturn V rockets and the procurement of new hardware called for sustained budgets of the mid-1960s level and that never happened.

But the story of the AAP plans is central to this book because it was the early designs toward an Extended Lunar Module, in development of scientific instruments which could be carried in an Apollo Service Module, in the more capable personal back packs being designed for longer work sessions and in the conceptual designs for mobile vehicles, that the expanded capabilities of the H and J flights were realised.

Yet by 1968 all hope of extending the lunar exploration phase through additional hardware had vanished but, with the goal being achieved with much less hardware than originally anticipated surplus spacecraft could feed directly into the utilisation of remaining vehicles, particularly the Lunar Module, for advanced exploration. During 1968 and early 1969, working to the revised flight schedule and a more realistic awareness of just how quickly the interim objectives could be met, NASA sustained interest in developing an Extended Lunar Module (ELM) for a series of more sophisticated objectives.

By this date NASA had abandoned AAP and was already planning a post-Apollo programme, a dramatic restructuring of goals and objectives which would incorporate winged, reusable shuttle vehicles for delivering payloads and astronauts to low Earth orbit, space stations for scientific research, nuclear propulsion for upper stages promising much reduced transit times to deep-space objectives and an expanded programme of planetary exploration with autonomous robots. All this was designed to sustain a manned space flight programme at a lower budget. With that in mind, the remaining Apollo hardware elements were precious assets for delivering to science what had been absent

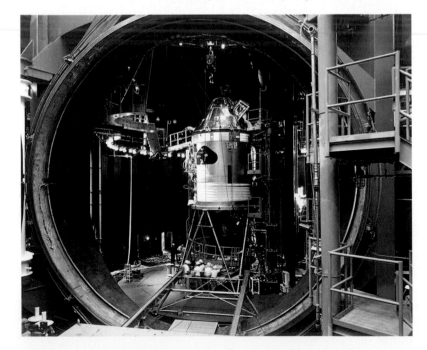

LEFT An Apollo test vehicle is delivered to an environmental simulation chamber at NASA's Manned Spacecraft Center where it will be put through a simulated exposure to the heat and vacuum of space. *(NASA)*

RIGHT Simulators at the Kennedy Space Center
where astronauts would rehearse procedures
for space flight, crucial assets before and during
flight, when corrective procedures for unexpected
problems could be evaluated before being
adopted during a mission. (NASA)

from the purely political goal set by Kennedy.

By August 1968, when the interim sequence
was modified to allow a lunar orbit flight
on only the second manned Apollo flight,
Owen Maynard had gathered up the evolving
strategies, most of which had been brewing
at the Manned Spacecraft Center where the
full potential was realised, and delivered a
supplementary rationale based on his alphabet-
steps. It included additional missions to the
initial landing which would fully exploit both
Apollo and the Lunar Module, as follows:

H Precision landing with a two-day stay on the
Moon, an Apollo Lunar Surface Experiment
Package (ALSEP) to deploy to the surface,
two EVAs and option of a non-free return
trajectory;

I An extended duration CSM-only flight in
lunar polar orbit with a comprehensive
photo-mapping objective utilising a Scientific
Instrument Module (SIM) in the Service Module;

J Utilising an Extended Lunar Module for
three-day stays on the lunar surface, three
EVAs, deployment of an ALSEP, use of a
Lunar Roving Vehicle for distant site surveys
and option of a non-free return trajectory for
reaching high latitudes.

The objective with the H and the J series was
to divide remaining and available missions
between the two, with the notional decision to
support at least four J-series flights. In truth, the
G mission engaging the first landing was a sub-
variant of the H-series, because the G mission
was cut back from a more ambitious flight plan
through reduced lunar surface procedures to
increase the engineering return without over-
stressing spacecraft or crew. At this date, of
course, nobody knew how the Lunar Module
would perform at the surface and whether
there would be issues regarding safety and
spacecraft survivability.

BELOW The AS-204 launch vehicle that would have been used to put the
manned Apollo 1 spacecraft in orbit was used to test the first Lunar Module
(LM-1) shown here at the Kennedy Space Center. (NASA)

2 Development flights

──(○)──────────────

Before advanced missions could be flown to the lunar surface, with a new capability for conducting science from orbit, a series of development flights was necessary to bring the hardware, the engineers and the flight controllers through one of the most intense periods of preparation NASA had experienced – five missions to demonstrate that it all worked.

OPPOSITE Seen by Schweickart as he conducts an EVA from the front hatch of the Lunar Module, Scott retrieves science patches from the docked Apollo spacecraft. Note the hand rails to facilitate access should the tunnel between the two spacecraft become blocked. *(NASA)*

C-1 Apollo 7

Launch date: 11 October 1968
Duration: 260hr 09min 03sec

The flight of Apollo 7 began almost 20 months after the first manned flight had been expected, and for which the Apollo 1 crew had been training. Flown by the original Apollo 1 back crew of Schirra, Eisele and Cunningham, it was a technical success in that it demonstrated the effectiveness of quality control and design modifications which had transformed the CSM into a fine and robust vehicle fit for deep

LEFT The Apollo 7 crew had been back-up to the crew lost in the Apollo 1 fire and were assigned to fly the significantly modified Block II spacecraft as Apollo 7 in October 1968. From left: Walt Cunningham, Wally Schirra and Don Eisele. *(NASA)*

BELOW Lift-off for Apollo 7 as the Saturn IB roars toward space from Launch Complex 34, as seen beyond rocket gantries at Cape Canaveral. *(NASA)*

space expeditions. The mission itself provided opportunities for conducting navigation exercises and was able to conduct some modest demonstration of turning around and moving back in on the top of the S-IVB stage from where, on a lunar-bound mission, the Lunar Module would be restrained before extraction after docking.

Another valued test was the ability to flex the capabilities of the Earth-based tracking and ranging system which was a vital component of spacecraft guidance and navigation, since the primary source of that information came from the ground with on-board capability used for aligning the platform, re-setting the platform in the event of a gimbal-lock or conducting course corrections and manoeuvre parameters in the event of a total loss of communication with Earth.

From an engineering standpoint the flight was an unqualified success, raising hopes that the first landing attempt could be made after three more test flights. Apollo 7 did more than that, for it opened the window to the flight of Apollo 8 on a newly planned C' mission to carry humans beyond the gravitational attraction of the Earth and into the grip of the Moon. Intelligence information indicated that the Russians were leading up to something special and circumlunar flights had already taken place, with biological experiments included on flights with unmanned variants of the Soyuz spacecraft under the designation Zond.

The circumlunar Zond programme was a desperate attempt to upstage the Americans by becoming the first to fly humans around the Moon – with no plan to enter lunar orbit – but the decree authorising such an attempt was not issued until 1967. Intelligence information about Soviet plans was remarkably accurate, with the certain knowledge that the first crewed Zond circumlunar flight was scheduled for 8 December 1968, in a last-ditch attempt to beat NASA to what was regarded as a very high propaganda prize. If Apollo 8 as a circumlunar mission was driven largely by the delay to the Lunar Module, the decision to go had a political nuance too.

The decision to fly Apollo 8 as a lunar-orbit mission is full of drama and the story of how a few senior managers convinced the NASA leadership to support this initiative has been told many times and is available elsewhere, but the

ABOVE Soon to become synonymous with dedication, teamwork and a no-nonsense approach to the role of Flight Director, Eugene (Gene) F. Kranz monitors the launch of Apollo 7 from Mission Control in Houston. *(NASA)*

BELOW One purpose of Apollo 7 was to move back in on the S-IVB stage and evaluate the ability of the Apollo Command and Service Modules to approach the Spacecraft Lunar Module Adapter (SLA) and move back to where, on a Moon mission, the LM would be located and from where it would be extracted. *(NASA)*

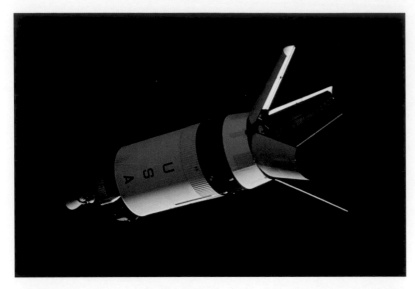

ABOVE The open petal-doors of the SLA were considered to be a potential hazard if one, or all, only partially opened and obstructed access to the Lunar Module. They were jettisoned when the Apollo spacecraft separated on future missions. *(NASA)*

BELOW The crew of Apollo 8 had originally been expected to fly a highly elliptical Earth orbit but found themselves going to orbit the Moon. From left: Frank Borman, Bill Anders and Jim Lovell. *(NASA)*

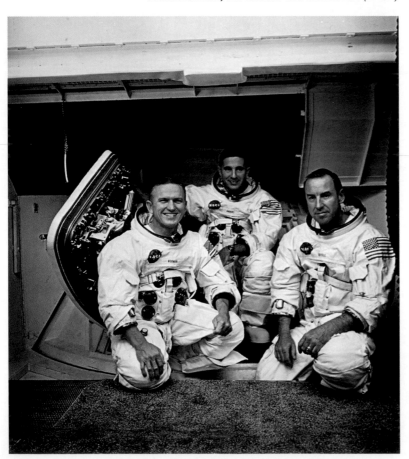

crew were solidly behind the change of mission objective and keen to demonstrate a highly significant step forward for the programme. For some astronauts the ability to outpace the Russians was a secondary consequence to a personal challenge as an extension of a career, but to Frank Borman there was a very conscious awareness of what Apollo had been set up to achieve and his loyalty to the political message and the drive to outflank Soviet plans was a highly motivating factor in his judgement about the balance between value and risk.

C' Apollo 8

Launch date: 21 December 1968
Duration: 147hr 00min 42sec

CSM-103 was the first fully loaded Apollo spacecraft to fly with a crew. The Apollo 7 loadings had been less due to the limited payload capability of the Saturn IB; CSM-103 having a mass of 63,531lb (28,818kg) versus 32,495lb (14,740kg) for CSM-101. There was nothing challenging about that, other than the new operating techniques that would be tested for the first time, one of which was about the requirement to slowly roll the spacecraft around its longitudinal axis to distribute thermal energy in the vacuum of space on the way to the Moon and on the way back.

Known as Passive Thermal Control (PTC), it embraced selective thermal coatings and colours for the two types of radiator mentioned earlier during descriptions of Block I and Block II spacecraft, and the water-glycol active control loops operating with cold-pipes and cold-trays. These were structures through which the coolant flowed on demand according to temperature levels and operating thermal bandwidths, physically attached to which were the systems that required active control.

Ideally, the spacecraft was placed so that the longitudinal axis was perpendicular to the Sun with the RCS thrusters used to set up a rotation rate of about 0.1°/sec, which would allow the spacecraft to make about one revolution every hour. Over time, the spacecraft began to wobble and prescribe an ever larger cone, as any rotating gyroscope will tend to do. If left uncorrected the CSM would point first the apex of the Command Module and then the SPS engine nozzle at the

RIGHT Apollo 8 was the first manned Apollo-Saturn V launch, the AS-503 launcher seen here framed with the fixed umbilical tower to the left and the Mobile Service Structure which was withdrawn before launch. *(NASA)*

back of the Service Module to the Sun. For future missions an automatic corrective adjustment to the coning effect was programmed in to the attitude orientation software.

Another aspect was the dynamic control of the free-return trajectory, a type of flight path which would automatically return the spacecraft on a figure-of-eight after passing around the western limb of the Moon within any discrepancies within the delta-velocity of the small attitude control and manoeuvring (reaction control) thrusters, or RCS units, in four quads on the Service Module.

The planned retrograde orbit had advantages in that the exchange of energy with the Moon provided a reduced ΔV for the Lunar Orbit Insertion (LOI) burn because its heliocentric velocity was lower. Mission rules dictated that before receiving a 'go' for LOI, the big SPS engine would have to be fired, its performance carefully monitored to approve its use for the all-important burn; from lunar orbit, a Trans-Earth Injection (TEI) burn could not be performed by the RCS thrusters. This was accommodated in a unique manner by deliberately displacing the trajectory away from the nominal and then using the SPS engine to place it on its final, and planned, flight path.

After separating from the S-IVB, an initial thruster firing at 4hr 45min 01sec conducted a 7.7ft/sec (2.35m/sec) velocity change which moved the flight trajectory to a projected lunar miss distance of 527ml (848km). This was deliberately set up so as to provide an opportunity for an SPS burn for the first time on this mission to verify acceptable engine performance before committing to lunar orbit.

RIGHT The flight to the Moon was given urgency by intelligence information that suggested the Russians were about to conduct a manned circumlunar Zond mission, the two spacecraft here compared for size. *(Eberhard Marx)*

That miss distance was also optimum for a free-return flight path should nothing further be done to the trajectory, returning the spacecraft to Earth within the capabilities of the small RCS thrusters. The burn came at an elapsed time of 10hr 59min 59.5sec, a 2.4sec burn changing velocity by 24.8ft/sec (7.56m/sec), moving the trajectory back for a fly-by at 76.3ml (122.8km).

The spacecraft had slowed to a velocity of 2,223mph (3,578kph) at an elapsed time of 55hr 38min, when it came within the gravitational attraction of the Moon, a point known as the equigravisphere. From there it began to speed up until passing around the far side of the Moon. A final tweak burn with the RCS thrusters at 60hr 59min 55sec changed velocity by 4.1ft/sec (1.25m/sec).

The LOI burn occurred at a height of 87ml (140km) above the surface of the Moon and an elapsed time of 69hr 8min 20sec, some 9min 35sec after disappearing from direct line-of-sight with Earth for the 35 minutes it took for Apollo 8 to traverse the far side. The SPS engine fired for 4min 6.9sec, reducing its speed by 2,043mph (3,287kph) and dropping Apollo 8 into an orbit of 69ml x 194ml (111km x 312km). A complete circumnavigation of the Moon took about two hours and on the second pass, on the far side, the SPS engine fired again at 73hr 35min 07sec

to circularise the orbit at 69ml x 70ml (111km x 113km). This was known as LOI-2.

Apollo 8 completed ten orbits of the Moon before firing the SPS engine a third time, at 89hr 19min 16sec for 3min 23.7sec, adding 2,400mph (3,862kph) to its velocity and shooting Apollo 8 out of lunar orbit and on to a trans-Earth trajectory. A detailed test objective was to demonstrate star-Earth landmark optical navigation but the accuracy of other navigation modes precluded the need for this to secure data for course corrections.

D-1 Apollo 9

Launch date: 3 March 1969
Duration: 241hr 00min 54sec

The Apollo 9 crew of McDivitt, Scott and Schweickart had originally been assigned to fly the second manned Apollo spacecraft, the first Block II configuration, as AS-205, then designated Apollo 2, but they were moved to the third manned Apollo mission after the fire put the Apollo 1 back-up crew on the replacement flight, re-designated Apollo 7. Then they moved further down the flight line, flying the third manned mission as Apollo 9 when the combined flight of the CSM and LM was replaced by Apollo 8.

Seat-shuffling was a not infrequent occurrence during this hectic period and would greatly influence the crew who landed on the Moon first. In reassigning prime and back-up crews for Apollo 9 to Apollo 8, that moved the back-up crew of Conrad, Gordon and Bean down a mission to Apollo 12 instead of Apollo 11, fortuitously putting Armstrong, Collins and Aldrin on the first landing attempt. There was no deliberate selection of the Apollo 11 crew through personality, as has been frequently indicated in several previous publications.

Scheduled to fly on 28 February 1969, the Apollo 9 crew were mildly affected by head colds which delayed the attempt for three days for suitable medication. It involved the first full complement of payload, with CSM-104 and LM-3 carried into space by Saturn V AS-504, on a shakedown flight for the Lunar Module which for this mission had been named Spider; the Apollo spacecraft had the call sign Gumdrop. Several changes were introduced on AS-504 including a more powerful S-II second stage, the output of the five J-2 engines being increased by 2.2% to generate a stage thrust of 1.15million lb (5,115.2kPa), while the empty stage itself was lightened by almost 3.9% to just 84,600lb (38,374kg). Weight saving decreased the dry mass of the S-IVB by 1.5% to 259,337lb (117,635kg).

The primary mission objective was to evaluate crew operations with the LM and fly it away from the CSM utilising a separation, rendezvous and docking procedure which had been worked out for a lunar mission. Critical operations involved the CSM detaching itself from the base of the Spacecraft Lunar Module Adapter (SLA), turning around, moving in to dock with the LM and extracting it from its stowed position. Several propulsive manoeuvres

ABOVE A view of the Earth, the first taken by humans showing it as an isolated sphere set in the blackness of space, a picture that would encourage environmental pressure groups to lobby for greater attention to its fragile ecosystem and diminishing resources. *(NASA)*

LEFT The Apollo 9 crew would be the first to conduct a dress rehearsal with the Apollo spacecraft and the Lunar Module, the first time the latter would be manned. From left: Jim McDivitt, Dave Scott and Rusty Schweickart. *(NASA)*

ABOVE The Lunar Module nestled inside the lower section of the SLA viewed after the Apollo spacecraft had separated and began moving back in to extract the LM, the door panels now having been jettisoned. *(NASA)*

BELOW The Lunar Module as seen from the Apollo spacecraft during free-flight operations that would take it far away from the only vehicle by which the astronauts could return home. *(NASA)*

with the LM were planned for independent flight and a range of engineering objectives involved the Apollo SPS engine as well as the LM descent and ascent propulsion systems.

The launch and ascent to orbit was conducted as planned and the S-IVB duly placed the two spacecraft, weighing 95,231lb (43,197kg), in orbit. Scheduled mission operations began with a short burn of the SPS to verify satisfactory operation but picked up on the second day in space with three SPS burns, the first time this engine had been fired in the docked configuration and with the much greater mass involved.

In this configuration the stack weighed 91,000lb (41,278kg) with a long moment arm against which to test the engine gimbal jacks maintaining centre of thrust through the centre of mass. This provided valuable data on the attitude excursions imposed by the thrusting SPS engine, the longest burn of the three being 4min 40sec which raised apogee to 316ml (503km), reducing the mass of the stack and shifting the longitude of the orbital ground track 10° east for better sunlight angles during the separation and rendezvous manoeuvres when Spider flew away from and returned to Gumdrop. The last of the three burns tested extreme gimbal angles on the SPS engine drive system and moved the orbit a further 1° east to test dynamic loads under a severe steering input.

The initial checkout of the LM took place on the third mission day for a full evaluation of Spider in operational configuration. A major event was the firing of the Descent Propulsion System (DPS) in the docked configuration. This was to simulate the employment of that motor in pushing a docked stack out of lunar orbit and on its way back to Earth. A variation of that was conducted on Apollo 13 when the DPS was used to restore a free-return trajectory and then to speed up the return journey after the nearly catastrophic loss of the Service Module on the way to the Moon. For this evaluation, on Apollo 9 the docked burn lasted 6min 12sec involving a variety of throttle settings.

The fourth day was assigned to a test of the Personal Life Support System (PLSS) backpack which was being tried out in space for the first time. The PLSS was an independent unit providing oxygen, thermal control to the suit, biomedical connections, and a communication system which would provide life-support functions for several hours during operations on the surface of the Moon. It was also a test of the A7L space suit which was considerably more developed than the G4C suit used by Eugene Cernan in June 1966 during the second EVA performed by a US astronaut. On that flight the rigidity of the suit and the inadequacy of the life support system made it impossible for Cernan to carry out his tasks and the spacewalk was aborted.

Planned as a contingency in the event that the intravehicular tunnel became unusable for getting the LM crew back in the Apollo spacecraft, it envisaged the use of handrails and handholds attached to the exterior of the two docked vehicles so that crewmembers could traverse between the open hatches. But the trial was abbreviated after Schweickart, the crewmember who had trained on these procedures, suffered from space sickness. Instead of making his way from one vehicle to another, he stood on the 'porch', the platform across which Moonwalkers would move from the front hatch of the LM to the ladder on the forward landing leg.

The following day was devoted to the vital demonstration of CSM/LM undocking, separation manoeuvres and rendezvous and re-docking operations, the most anticipated events on the mission. Spider was in free flight for more than six hours. At a range of 84ml (135km) behind Gumdrop the Descent Stage was jettisoned after McDivitt and Schweickart had switched power and consumables supply to the Ascent Stage, and the LM's thrusters were fired for 32 seconds, both events occurring at 96hr 16min 06sec in a Coelliptic Sequence Initiation (CSI) manoeuvre, stopping the drift away from Gumdrop as it dropped Spider down to a lower orbit than the Apollo spacecraft, halfway to which the separation distance had increased to 115ml (185km).

To conduct a rendezvous with Gumdrop, Spider had to increase its height and in so doing slow down in a synchronised sequence known as Terminal Phase Initiation (TPI). This latter sequence represented the relative positions of an Apollo spacecraft and an LM in orbit around the Moon after the Ascent Stage had lifted off to meet the CSM. TPI began

at 97hr 57min 59sec and slowly closed the distance up to Gumdrop through a series of two course corrections, all the while reducing relative velocity and converging on the CSM. Docking was achieved at 99hr 02min 26sec using the Crewman Optical Alignment Sight (COAS), an aid to aligning the spacecraft using a small telescope device to visually align the sight on a T-bar on the top of the Lunar Module.

After McDivitt and Schweickart transferred back into Gumdrop, Spider was configured for semi-autonomous post-separation activity and at 101hr 22min 45sec the Ascent Stage was released, followed at 101hr 53min 15sec by the second firing of the Ascent engine to oxidiser depletion at 6min 02sec which placed that stage in an orbit of 4,328ml x 145.7ml (6,965km x 234.5km). It would continue to orbit the Earth in this highly eccentric path until it re-entered the atmosphere and burned up on 23 October 1981; the Descent Stage would succumb to a similar fate on 22 March 1969.

The crew now had almost six days of orbital activity involving landmark tracking, navigation checks and two more SPS burns for data on different orbital set-ups. And there were several Earth photography tasks, including multispectral photography using the

ABOVE McDivitt and Schweickart in the Lunar Module view the Apollo spacecraft manned by Scott. *(NASA)*

multispectral camera system which consisted of four 500-EL Hasselblad cameras each fitted with an 80mm focal length lens with separate filter combinations.

Initial evaluation of the Apollo 9 flight cleared the LM for lunar operations, effectively satisfying the requirements set for the D-series objectives. There were, however, some changes to the hardware which would be introduced beginning with LM-5 assigned to Apollo 11, as this was the earliest that they could be made. Most apparent was the requirement for attaching exhaust plume deflectors to the Descent Stage to protect the structure from flame damage, something noted visually from the Earth orbit mission. All of this on the way to clearing the programme for the next step up.

F-1 Apollo 10

Launch date: 18 May 1969
Duration: 192hr 03min 23sec

The flight plan for Apollo 10 incorporated significant steps beyond anything tried before and for that it would be necessary to take the LM down to the point where, on a landing mission, the crew would start down to the surface with Powered Descent Initiation (PDI) 50,000ft (15,240m) above the surface. To some observers it seemed a waste to send Apollo 10 all the way and not let it land. But apart from the need for more engineering data, in order to provide an accurate simulation of the weight an Ascent Stage would have after firing its main engine from the surface of the Moon back into orbit, Apollo 10's LM would not carry sufficient propellant to get off the surface and back up even if it did land.

Those same individuals who claim that this was so that the Apollo 10 crew would be discouraged from making a quick dash to the surface fail to understand the engineering imperatives driving not only this mission but the entire programme, not to mention the professionalism of the astronauts. However, it didn't help when an ebullient Gene Cernan joked on more than one occasion that this was in fact the real reason for virtually empty tanks!

There was also a need for greater understanding of the lunar ephemerides, driven by the essential requirement for guidance and navigational inputs that would reduce the magnitude of inaccuracy on specific locations to at least 1.15ml (1.8km). Data from Lunar Orbiter, an unmanned spacecraft that mapped the Moon, revealed areas below the surface where greater concentrations of dense and massive material pulled and tugged at the orbital path, changing it over several revolutions. These areas were known as mascons (mass concentrations), with other areas below the mean average mass density of the Moon known as minicons. For Moon landing missions, knowing the orbital parameters was essential.

Defined by these conditions, known as the mascon/minicon problem, the accuracy from Lunar Orbiter data was 23ml (37km). Incorporating navigational plots from Apollo 8, the resolution was still only 5.7ml (9.1km) and a landing attempt conducted under those circumstances could take the LM that distance off course in the predicted trajectory calculated to reach the carefully selected landing site.

For this full-dress rehearsal, Stafford (Gemini VI-A and IX-A) would command a crew consisting of Young (Gemini 3 and X) and Cernan (Gemini IX-A). Training had been tempered by subtle changes to trajectories and flight plans and revised operational procedures had been evaluated and tested on preceding flights.

Changes to the AS-505 launch vehicle resulted in this being the heaviest Saturn V launched to date, placing the heaviest payload yet in Earth orbit (98,273lb/44,576kg). This

BELOW Apollo 10 was the second Moon mission, crewed by three veteran astronauts. From left: Gene Cernan, Tom Stafford and John Young. *(NASA)*

would also produce a record mass in lunar orbit of 69,429lb (31,493kg) for the combined CSM/LM configuration.

The Apollo spacecraft (CSM-106) was largely similar to CSM-104 (Apollo 9) with only minor changes. Some significant changes were made to the Lunar Module (LM-4), with increased strength in the structural webs of the Descent Stage through thickening by 0.015in (0.381mm) and bonded doublers fitted to the upper deck webs. In addition, two of the four batteries previously in quad 4 were moved to quad 1. Changes were also made to the composite structure of the thermal blankets and to the thermal shielding protecting the exterior from plume impingement from the RCS thrusters on the Service Module in the docked configuration.

As with Apollo 9, two spacecraft in operation at the same time necessitated call signs and the Apollo 10 crew chose Charlie Brown for the CSM and Snoopy for the LM, both taken from the Charles Schulz cartoon strip. These things were heavily influenced by the judgemental eye of Julian Scheer, head of public affairs. Appalled at the frivolity, he quickly got the approval of senior management for a more sober use of call signs and after this mission they took on a more conservative application of names, usually great sailing ships, names associated with discovery and great deeds or national symbols.

Following launch on 18 May 1969 at 11.49pm local time, the lift-off, ascent and Earth orbit phase of the Apollo 10 flight was similar to that performed by the previous four Saturn V launches but on this flight, in an attempt to reduce the vertical oscillations still felt on the first stage, it was planned that the centre engine would be shut down 25 seconds before the four outer engines, a sequence phasing which would remain standard procedure on subsequent vehicles. Moreover, the centre J-2 engine on the S-II stage was also shut down early, 92sec before mainstage cut-off, to prevent low-frequency oscillations observed on the two previous Saturn V launches.

TLI placed the stack on a free-return flight path with a closest approach to the Moon of 1,044ml (1,680km), a figure which would be modified later. Showing separation, turnaround and docking with the LM, the crew in Charlie Brown gave Earth viewers the first live colour

LEFT The Lunar Module being lowered into the SLA which will be attached to the top of the S-IVB, the third stage of the Saturn V. (NASA)

TV pictures from space, starting transmission with the configuration 3,500ml (5,600km) distant and ending 22 minutes later 9,420ml (15,150km) away. This was followed by a further transmission 5hr 6min into the mission from a distance of 24,160ml (38,873km) which lasted 12 minutes.

The standard evasive manoeuvre occurred at 4hr 38min 47sec when the SPS engine was fired for 2.5sec for a velocity change of 18.7ft/sec (5.7m/sec), shifting the point of closest approach to 358ml (576km). The purpose of that built-in change was based on a requirement to have the inclination of the

BELOW Cernan in the foreground waits for Stafford to enter the Apollo 10 Command Module prior to launch. (NASA)

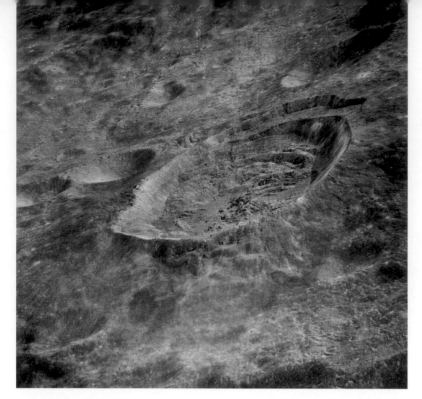

ABOVE Necho Crater on the far side of the Moon, an impact depression about 20ml (32km) across, as viewed by Apollo 10. (NASA)

the way in which the rotating mass would begin to cone and the use of the thrusters to stabilise the axis of rotation caused problems, in that the frequency of firing the tiny rocket motors caused disturbance in the axial pointing angle.

The LOI-1 burn was conducted satisfactorily at 75hr 55min 54sec lasting 5min 56sec, much longer than the Apollo 8 insertion burn due to the very different weights at this point in the mission: 93,318lb (43,000kg) for Apollo 10 versus 63,500lb (28,800kg) for Apollo 8. The achieved orbit was 195.6ml (315km) by 69.3ml (111.5km), very close to the pre-planned value. The LOI-2 circularisation burn occurred with ignition of the SPS engine at 80hr 25min 08sec placing the configuration in an orbit of 70.2ml (113km) by 68.1ml (109.5km).

The crew entered the Lunar Module for the first time at around 82hr elapsed time when equipment from Charlie Brown was transferred to Snoopy ready for the long day of independent activity, before a rest period and preparations for the big day when the initial stages of a landing profile would be tested around the Moon. The dramatic achievements of Apollo 8 in being the first manned spacecraft to achieve lunar orbit and the orbital ballet conducted by the first manned flight of the Lunar Module on Apollo 9, seemed to eclipse in public perception the dramatic importance of this particular flight.

Undocking of the two spacecraft occurred at 98hr 11min 57sec, the standard profile for a lunar landing flight and preceding the PDI burn where the Descent Propulsion System would fire continuously, braking the velocity of the LM and taking it down to the surface. The 27sec Descent Orbit Insertion burn with the LM came at 99hr 46min 01sec on the far side of the Moon, placing Snoopy in an orbit of 70ml (112.6km) by 9.8ml (15.78km) so that pericynthion would occur on the near side at 15° east longitude and east of the landing site planned for the initial landing. Being on an elliptical path which would carry them close to the lunar surface, the LM speeded up and was now 185ml (298km) ahead of Charlie Brown above. At pericynthion, the LM was 50,497ft (15,392m) above the radius of the landing site and by the time they arrived overhead at LS-2 they were at 56,783ft (17,307m) and climbing slightly out of the low point in the orbit.

lunar orbit so that the approach azimuth to the landing site (LS-2) would be close to that anticipated for the first landing attempt. The final cislunar trajectory correction was performed at 26hr 32min 56sec for a 6.67 burn of the SPS engine, changing velocity by 49.2ft/sec (15m/sec), shifting the closest approach to a mere 70ml (112.6km) above the lunar surface.

One of the primary activities during the coast to the Moon was to place the docked assembly into a PTC 'barbecue' mode as described for the Apollo 8 mission but the different centre of mass for the two spacecraft and the very different weight distribution compared to the lone CSM on Apollo 8 brought some surprises. It had not been possible to completely model

RIGHT The Ascent Stage returns to the Apollo spacecraft after several hours spent in manoeuvres that rehearse the procedures a future Moon landing would undergo. This precursor flight exposed several areas where improvements in procedure, tracking and flight management were necessary. (NASA)

But they wanted to get behind as well as below the CSM so that they could simulate the position a crew would be in returning from the Moon's surface. To achieve that they needed to conduct a further burn to put them in a wide, looping flight path moving far above the CSM, slowing down in the process and coming back down again to the same relative distance behind the mothership as a crew coming up from the Moon. This 40sec phasing burn took place at 100hr 58min 25sec, placing Snoopy in a highly elliptical path with an apocynthion of 218ml (350km). As the LM flew that higher path it slowed and that allowed the CSM to get ahead.

This arching trajectory lasted 1hr 47min during which time Stafford and Cernan configured Snoopy for simulating the rendezvous operations all future missions would follow, a sequence which had been closely rehearsed on Apollo 9. But that had been in Earth orbit where the times, revolutions and manoeuvres were compromised by the greater mass of the home planet. The idea here was to separate the Ascent Stage from the Descent Stage to more closely simulate the situation when a crew was returning from the surface (which is why, as related earlier, that stage was only partially filled with propellant).

Staging occurred at 102hr 45min 17sec and the 'insertion' burn took place less than eight minutes later, a firing of the Ascent Propulsion System for 15.6sec placing the stage in a 52.1ml x 12.9ml (83.8km x 20.7km) path. In this orbit, Snoopy would slowly catch up with Charlie Brown through a series of intermediate manoeuvres and would eventually reach a relative position behind and below the CSM so that it could start the rendezvous sequence beginning with TPI (see Apollo 9). At apocynthion, just over 50min after the insertion burn, the crew used Snoopy's RCS thrusters to perform the Coelliptic Sequence Initiation burn, changing to an orbit of 54.3ml x 48.1ml (87.4km x 77.4km). The TPI burn took place at 105hr 22min 55sec, preceding docking which was achieved at 106hr 22min 02sec.

After the crew rejoined Young in Charlie Brown, and equipment was transferred from the CSM to the Ascent Stage, what remained of Snoopy was jettisoned at 108hr 24min 36sec, followed 19min later by the CSM conducting a separation burn to

clear space for the APS engine to fire to depletion. That occurred 9min later for a duration of 4min 48.9sec, placing it into an unknown heliocentric orbit. After a total separation time of 8hr 10min 5sec, all three crewmembers had accomplished a highly significant set of objectives that would leave nothing standing in the way of a lunar landing attempt.

Landmark tracking activity occupied the crew for the last day in lunar orbit before the TEI burn at 137hr 36min 29sec, placing Apollo 10 on a fast track to Earth, crossing the distance in around 54 hours with recovery by the USS *Princeton* just 39 minutes after landing.

G-1 Apollo 11

Launch Date: 16 July 1969
Duration: 195hr 18min 35sec

The G mission in Owen Maynard's sequence of alphabet-steps was the one which would finally qualify the hardware for the several H-series lunar exploration flights to come. That capability would be realised before the demonstration and qualification steps (A–G) had been achieved.

It did not take long for a post-flight analysis with the F-mission to decide to progress toward the G-mission and the first landing attempt. This was an experienced crew consisting of Commander Neil Armstrong (Gemini VIII); Command Module Pilot Michael Collins (Gemini X); and Lunar Module Pilot Edwin 'Buzz' Aldrin

ABOVE Apollo 10 comes home and lands close to the USS *Princeton* for recovery by SH-3D helicopter from the carrier. *(NASA)*

ABOVE The Apollo 11 astronauts (from left) Neil Armstrong, Mike Collins and Buzz Aldrin, selected to fly the final missions development flight to qualify the programme to support longer duration missions and extended activities on the surface. *(NASA)*

(Gemini XII). In addition to being in command of the overall mission, Armstrong was the man who would 'fly' the LM down to the surface, Aldrin being essentially the systems engineer who would not be in control of either the CSM or the LM at any point, his job being to monitor equipment and readouts and, in the LM, to provide information to Armstrong.

The crew for Apollo 11 had been announced on 20 November 1967, but at that time Jim Lovell was in the slot for CMP until he was replaced by Collins, after he had been replaced by Lovell on Apollo 8. Collins had a bone spur which had to be removed and so he recycled

RIGHT No significant picture exists of Armstrong on the lunar surface but this image of Aldrin became emblematic of the first time humans set foot upon the surface of another world, July 1969. *(NASA)*

back into the roster, placing him on Apollo 11 after Lovell stood in for Collins on Apollo 8. Despite having flown so recently, Lovell was offered to Armstrong as replacement for Aldrin, whom many considered to be difficult to work with, but Armstrong declined on the basis that he had never experienced issues with Aldrin. The back-up crew consisted of Lovell, Anders and Haise, with a support crew comprising Mattingly, Evans and Pogue.

Surface activity would consist of deploying the Early Apollo Scientific Experiments Package (EASEP) and conducting a documented sample collection, with most of the photography conducted by Armstrong. The only tools available for what was itself considered to be a development and verification activity before serious geological work could begin on later missions, included a large scoop for obtaining loose surface material, an extension handle to fit the large scoop, a core tube and hammer and tongs for collecting small rocks. A gnomon was carried, to be set down on the surface, for reference to the local gravitational vertical and two Sample Return Containers (SRCs) for sealing up to 130lb (59kg) of surface material in sample bags along with the tools.

Lift-off for Apollo 11 occurred at the start of the launch window at the first launch opportunity on 16 July 1969 at 09:32hr local time and all of the basic flight milestones were followed as demonstrated by Apollo 10. There were no significant departures from those flight plan events and the TLI burn was performed so that the docked vehicles would pass the Moon at an altitude of 808ml (1,300km). An evasive burn to avoid the trajectory of the spent S-IVB was conducted at 4hr 40min 01sec, a 3.4sec SPS burn to move perilune to 207ml (333km). Only one of the four MCC manoeuvres was conducted, performed with the SPS engine on Columbia for a 20.9ft/sec (6.43m/sec) velocity change at 26hr 44min 58sec, shifting perilune further in at 72.3ml (116km).

The story of Apollo 11 has been told many times and suffice to outline here that the two Lunar Orbit Insertion burns and the Descent Orbit Insertion burn by Eagle went off as planned and as had been simulated by Apollo 10. Powered Descent Initiation occurred at 102hr 33min 05sec. The four parts of the powered descent to the surface began with the braking phase, providing deceleration at an efficient rate for reducing orbital velocity and bringing the perilune down to a subsurface point, notionally lasting 8min 26sec.

Then the final approach phase would last 1min 40sec starting at 7,500ft (2,286m), a position known as High Gate, to allow for pilot visibility of the landing area. It terminated at Low Gate, an altitude of around 500ft (152m), for commencement of the landing phase which would last around 1min 30sec for manual takeover incorporating a vertical descent phase from a height of about 65ft (20m) and continuing down to the surface, a total powered descent sequence nominally lasting 11min 36sec.

Touchdown occurred at 12min 35sec, ending a safe and controlled descent. Due to a combination of navigational and procedural errors the LM landed 3.45ml (5.55km) long and very nearly in unacceptable terrain close to a depression known as West Crater. Manual override allowed Armstrong to move the landing 1,100ft (335m) downrange and 400ft (122m) south-west of where the guidance system under P-64/P-65 computer control was taking them. Had it proceeded on auto-guidance to the location it believed to be the correct one, the LM would either have been destroyed or an abort would have been required to save the crew.

Eagle remained on the surface of the Moon for 21hr 36min 21sec with Armstrong and Aldrin having conducted an EVA lasting 2hr 31min 40sec, deploying a solar-powered seismometer, a laser-ranging retroreflector and solar wind composition experiment. After ascent from the surface, the template of activities involving rendezvous and docking and the return to Earth with TEI followed the Apollo 10 mission sequence and completed the development phase of Apollo, realising the political objective of landing on the lunar surface before the end of the decade and releasing the remaining hardware for the scientific exploration of the Moon.

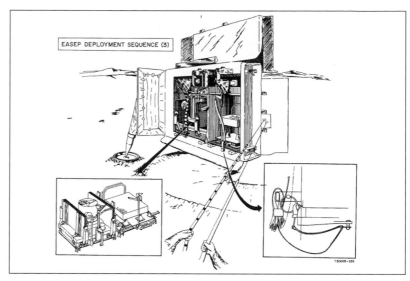

ABOVE The Early Apollo Science Experiments Package (EASAP) was contained in the Scientific Equipment (SEQ) bay on the Lunar Module Descent Stage and removed by a system of lanyards that opened the doors to allow access. *(NASA)*

LEFT Deployment of the passive seismic experiment (PSE) (foreground) and the laser ranging retro-reflector (LRRR), with Aldrin looking back at the LM. *(NASA)*

3 The operational missions

═══◯═══════════════════════

With completion of the five development flights the Apollo programme was left with sufficient hardware for a further nine Moon landings and this prompted a more concerted effort at determining just where to go on the lunar surface. Having promoted Apollo 11 as the last of the engineering development flights it is important to emphasise the evolving nature of the programme, now intent on exploration.

OPPOSITE Dramatic skies as AS-506 thunders toward an encounter with a lightning strike which will knock out almost all the systems in the Apollo spacecraft but leave the computers in the launch vehicle's Instrument Unit unscathed. *(NASA)*

mission staying for less than 22 hours, with follow-on missions carrying an ALSEP array for deployment during a surface stay time of up to 36 hours.

The four Phase 2 missions were projected for the period 1972–1973 and supported operations with an Extended Lunar Module (ELM) for three-day stays on the surface and three EVAs. They anticipated use of a Lunar Flyer, a rocket-powered pack capable of transporting up to two people across the lunar surface to access distant points. The emphasis in this phase was to have been on geological surveys over greater distances, providing an opportunity to access different geologic units. Several sites were proposed, including the Fra Mauro formation which would in fact be the destination for the aborted Apollo 13 and visited by Apollo 14.

Phase 3 missions would incorporate a range of objectives including the 28-day lunar polar orbit flight of a lone CSM for extensive and highly detailed photographic mapping, the I-series mission as shown in Owen Maynard's alphabet of increasingly sophisticated stages in advanced lunar exploration. It was suggested that the I-mission observation module, carrying a battery of advanced cameras, would be left behind in lunar orbit for autonomous operation over a further, extended period. Phase 4 was envisaged as two Lunar Surface Rendezvous and Exploration Missions, each consisting of dual launches in 1975–1976. These missions were never integrated into operational planning and extended into the domain of the Apollo Applications Program (AAP) and are best described elsewhere.

Just as each spacecraft was virtually hand built, incorporating unique features in capability and operability, it is equally important to discriminate between the standard Lunar Module and the Extended Lunar Module (ELM), which never did get that name but which here in this context serves to identify the second-generation LM capable of more advanced and ambitious objectives. The ELM would serve the requirements of the J-series missions but even with the operational H-series there were flight-by-flight changes to refine, improve and generally expand the capabilities of the basic spacecraft.

ABOVE The crew selected for the second landing on the Moon included (from left) Pete Conrad, Dick Gordon and Al Bean, the first science mission planned for full operational capability. The preceding six development flights had provided refined and updated procedures which would serve as a fitting prelude to extensive geological surveys of selected sites. *(NASA)*

From 1962 Bellcomm played a major role in defining the scope and expansion of the Apollo programme and served to define the way extended missions would proceed. It was primarily due to Noel Hinners, D. James and F. Schmidt that NASA was directed toward a permanent programme of lunar exploration, truncated in reality by political decisions. But the general direction of that effort would shape the way the actual landings were focused.

A year after the Apollo fire, in January 1968 they laid out a sequential plan envisaging 12 lunar landings spread across four separate phases. Executed between 1969 and 1971, Phase 1 would have embraced the final engineering mission on a free-return trajectory, one which greatly limited site options and supported four additional landings. The relevant spacecraft would be capable of carrying 300lb (136kg) to the lunar surface, only the first

But the Apollo spacecraft too had a 'Block III' adaptation, into a lunar orbiting science station for the J-series which would significantly increase the workload of the Command Module Pilot, about which more later on. The integration of orbital and surface science greatly expanded the opportunity for using the human space flight programme to integrate searching questions involving fundamental tasks of exploration and discovery. This alone did a tremendous amount to recruit scientists in support of the Apollo programme, where once they had been opposed to human space flight on the basis that it was sucking funds from unmanned activities.

In practical planning for future missions in the H and J series, NASA stood down the next mission after Apollo 11 from its planned September launch slot and this greatly relieved personnel from what had been an arduous

and intensive push to get the first man on the Moon. There was a need to scrutinise every detail of those development flights, especially the G-series mission of Apollo 11, and work through any engineering changes, any technical modifications to hardware and software and to assess the backlog of equipment for what were now envisaged as four H-series missions utilising the full potential of the hardware and the five J-series flights with the ELM and the developed Apollo CSM.

Bridging the gap between full qualification of the hardware for expanded operations was the requirement to tidy up the landing procedure and this fell back on to the global network of tracking stations and the Deep Space Network in particular, for supporting the remaining missions. Because a lot of the lunar sites the scientists wanted to get to required accurate landing, it was essential to clear up the issue

ABOVE Hardware stacking up at the Kennedy Space Center, indicated by this view of Apollo 12 hardware in the foreground with Apollo 11 spacecraft behind. With no certainty as to whether Apollo 11 would be able to land on the first attempt, Apollo 12 was ready to fly in September 1969. *(NASA)*

RIGHT The plaque for
Apollo 12 attached,
like all other missions,
to the forward landing
leg which supported
the ladder down which
the astronauts would
reach the surface.
(NASA)

H-1 Apollo 12

Launch Date: 14 November 1969
Duration: 244hr 36min 24sec

In essence, the need to demonstrate a pin-point touchdown became the final engineering refinement, an enabler for everything that was to follow. The crew had already been selected and consisted of Charles 'Pete' Conrad, Richard 'Dick' Gordon and Alan 'Al' Bean. An all-Navy crew, Conrad had flown on Gemini V and Gemini XI and Gordon, with Conrad, on Gemini XI, but Bean was making his first flight; only Gordon was making his last flight. Conrad and Bean would go down to the lunar surface in LM-6, which they named Intrepid, while Gordon would fly CSM-108, named Yankee Clipper.

The back-up crew for this mission comprised David 'Dave' R. Scott, Alfred 'Al' M. Worden and James 'Jim' B. Irwin. These men would form the crew for the first J-series mission, but at this time were expected to fly the H-4 mission had the original sequence been adhered to. In descending order of priority, the prime objectives were to demonstrate a pin-point landing, conduct further sample collection from a Maria region, deploy the ALSEP-1 scientific array, further expand the ability to work on the lunar surface and obtain additional photographs of future landing sites.

regarding pin-point touchdown within a few hundred metres at most of the desired spot.

Moreover, modified descent profiles were required, adapted through minor tweaks to the software and the computer programmes as well as refined simulator work for the astronauts, to drop down in mountainous areas rather than the flat Maria consisting of those expansive dried lava flood basins. This brought an integration of new and more advanced procedures for organising and operating the EVAs which would, in the fulfilment of the J-series missions, call for three full working days in succession out on the lunar surface.

For Apollo 12 the selected site was LS-7, situated at 2.94°S latitude by 23.45°W longitude, approximately 955ml (1,537km) west of the Apollo 11 site (LS-2). It was located in the Ocean of Storms and had an interesting attraction, to both verify an accurate touchdown and provide access to an existing lander. It was the site of the Surveyor III spacecraft, the second of NASA's soft-landers, which touched down on the inside rim of a shallow crater on 20 April 1967. The objective was to have Conrad and Bean land close by, the centre of the landing ellipse being 1,118ft (340m) from Surveyor III, walk to the spacecraft and retrieve sections of the tubular framing as well as the TV camera and return them to Earth.

There was much they could tell about the nature of materials left on the lunar surface for extended periods, in this case for 31 months. More still could be learned about the potential for biological activity on the lunar surface.

RIGHT The mission
badge for Apollo
12 showing Yankee
Clipper with Intrepid in
tow. (NASA)

Although assembled in a clean-room to protect delicate parts from dust, Surveyor III had not been sterilised and there was an open question as to whether microbes or any other form of biological activity could sustain life in such a vacuum and under solar radiation for most of its time on the Moon's surface.

But visiting Surveyor III would form an objective for the second EVA, the first being committed to sample collection and deployment of the ALSEP-1 array, activities which would take considerably more time than that for EASEP on Apollo 11. In fact, the site itself was considered geologically interesting because the Maria region was different to that at the Apollo 11 site and appeared younger. It also afforded the added advantage of possibly collecting ejecta material from the crater Copernicus, some 230ml (370km) to the north of the landing site, which would help to nail the age of this distinct feature on the lunar landscape.

Selection of the landing site

Selection of the site took place on 10 July 1969, from a list of 10 locations submitted to the Office of Manned Space Flight by the Group for Lunar Exploration Planning (GLEP), which included Noel Hinners, director of science and applications at the Manned Spacecraft Center (now the Johnson Space Center) and Farouk El-Baz, the supervisor for lunar science planning and operations at Bellcomm. Quickly, site selection was now switching from engineering verification to a meaningful survey for geologists. Lunar science was becoming an attractive field for research and the anticipation of samples back from several pre-selected sites attracted a new generation of scientists and invigorated those who never imagined they would ever have a role to play in getting rocks and soil back to Earth.

In the month before Apollo 11, assuming it was successful, the preferred site for Apollo 12 had been LS-5, a back-up site for the first lunar landing attempt but a bland and uninteresting place. But that was to change in favour of a previously low priority target. Getting there had

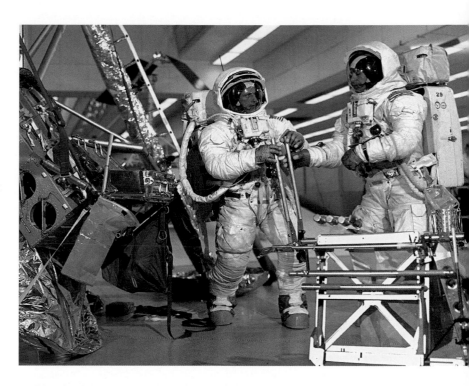

ABOVE Wearing training suits, Conrad (left) and Bean rehearse procedures for removing tools from the Modularised Equipment Stowage Assembly (MESA) to the right of the forwards landing leg in Descent Stage quad No 4. Note the Apollo Lunar Hand Tool Carrier (ALHTC), new for Apollo 12 and subsequent flights. *(NASA)*

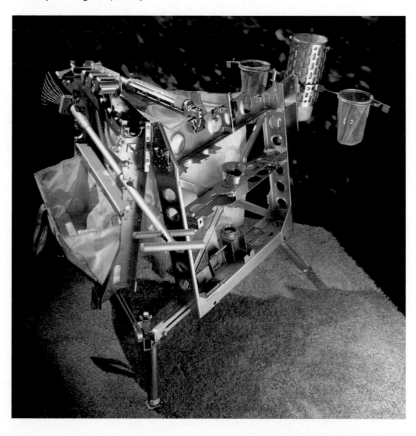

RIGHT The ALHTC on which a variety of tools, core tubes and sample bags could be attached and moved around on the surface. *(NASA)*

been a product of unforeseen circumstances added to concern about not putting the astronauts of a future mission in danger; it had been a big enough challenge putting Armstrong and Aldrin on the Moon without the added complexity of accessing a difficult site.

The former Surveyor programme manager Benjamin Milwitzky had written to Lee Scherer, then the director of the Apollo Lunar Exploration Office, and argued for a landing at the Surveyor III site, citing the value for accessing information about the condition of materials placed on the Moon more than two years earlier. The group also knew that, with improved photographs from Apollo 10 providing added value supporting the Apollo 11 site, the trajectory for accessing the Surveyor site would provide good 'bootstrap' photography of the Fra Mauro site for Apollo 13. But the Fra Mauro site was dangerous and the inability to get Apollo 11 near its intended landing site advised a conservative approach, which is why the LS-5 site was selected.

With the scientists anxious to get to the really exciting sites, calling for precision landing in rough terrain flanked by high mountains and certain dangers, the inability of Apollo 11 to set down where planned reopened the justification for putting Apollo 12 down at a safe and accessible 'test' site for demonstrating a pin-point touchdown. It was this imperative that drove management to assign the Surveyor III spacecraft as the target, designating it LS-7. But it was not all that easy knowing precisely where the spacecraft was on the lunar surface, as tracking the spacecraft had not provided the exact spot with sufficient accuracy to plan a landing within walking distance of the Lunar Module.

It was a largely self-taught British amateur astronomer who discovered precisely where it was. Ewen Whitaker worked with the eminent Gerard Kuiper on the US Army's lunar atlas of the late 1950s, when under Wernher von Braun there had been plans for the Army to support Moon landings using Saturn rockets. Whitaker

BELOW A training session which displays the MESA with open Sample Return Container boxes with open lids sample bags and the ALHTC in the foreground. The Surveyor III mock-up can be seen in the background. *(NASA)*

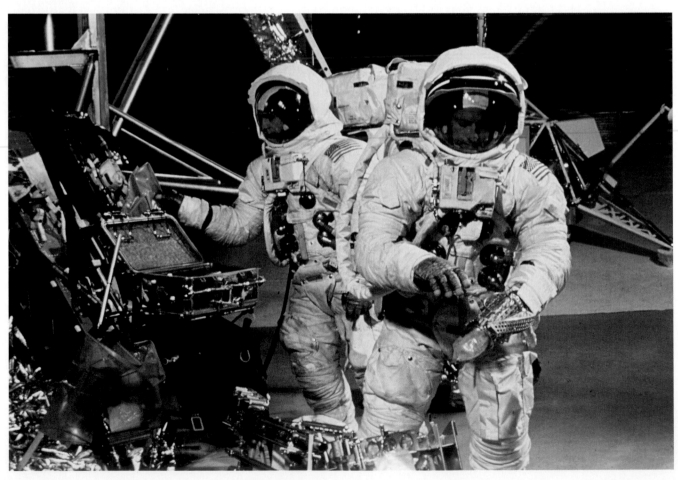

quickly gained a reputation for photographic analysis, selecting the site as the target for Ranger 4. Working with noted geologist Eugene Shoemaker, Whitaker applied his experience with finding the precise location of Surveyor I to search for Surveyor III.

The photographs from Surveyor III sitting on the surface showed a crater anywhere between 100–1,000ft (30–300m) across. Since only Lunar Orbiter III had provided images of the general area the resolution required examination of craters measuring 0.118–1.18in (3–30mm) across on photograph H154 from that mission. Relieved to find that the Sun angle and the shadow length of H-154 was about that at the time of the Surveyor III images, Whitaker used a small reading lens and a low-power telescope eyepiece to work his way across every crater on that picture!

After a self-assessed 26 hours of scrutinising every depression and crater, Whitaker had the precise location fixed. And that was enough to give approval for the Site 7 landing. The mock-up for Surveyor was brought over from Hughes Aircraft, the spacecraft's manufacturer, and Conrad and Bean began training with it, the replica situated at the same tilt angle

calculated from the photographs it had taken on the lunar surface. It seemed to some that the configuration of the craters around the landing site were set out like the shape of a snowman – so that was what the area was named and how it would be referred to in the air-to-ground communication.

The requirement for pin-point landing at

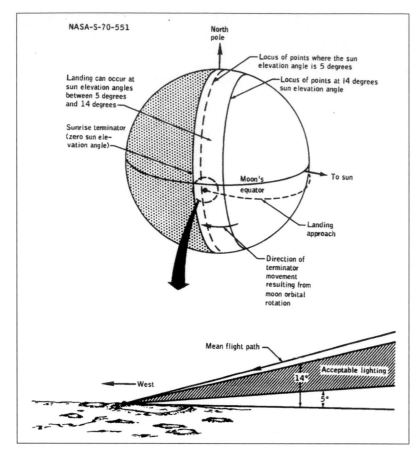

NASA-S-70-551

North pole

Locus of points where the sun elevation angle is 5 degrees

Locus of points at 14 degrees sun elevation angle

Landing can occur at sun elevation angles between 5 degrees and 14 degrees

Sunrise terminator (zero sun elevation angle)

Moon's equator

To sun

Landing approach

Direction of terminator movement resulting from moon orbital rotation

Mean flight path

Acceptable lighting

West

14°

5°

ABOVE The Sun elevation angle during descent and touchdown was a critical parameter for determining the launch window. This graphic indicates the allowable boundaries for observing features on the airless Moon where light is not scattered and shadowed areas are black. *(NASA)*

this more westerly site played well into the hands of engineering itself, providing a greater distance, and time, for tracking the LM after it came around the eastern limb of the Moon and that would ensure a more accurate powered descent phase. It was also a site without elevation obstacles and apparently devoid of deep craters, with less than 2° slopes on approach to ensure stable returns from the LM's landing radar signals. This site did require the spacecraft to approach the Moon on a non-free return trajectory but initially the flight path imparted by the S-IVB would be a free return type, the Apollo SPS engine being employed at MCC-2 to shift it to a higher elevation on approach to Lunar Orbit Insertion. It would be the first time a hybrid trajectory had been used.

A major aspect of the H-1 mission was the use of photography to observe distinct and unique phenomena to benefit selenology in general and to provide images for future site selection. The equipment carried by Apollo 12 included a 70mm Hasselblad EL (exclusive to the CSM), two 70mm Hasselblad data cameras, two 16mm Maurer data acquisition cameras

(one each in the CSM and the LM), one 35mm lunar surface close-up camera (carried in the LM MESA) and a four-camera multispectral experiment camera using Hasselblad ELs (for the CSM).

The flight

The launch of Apollo 12 occurred at 11:22:00am local time on 14 November 1969. The ascent phase on the S-IC first stage was nominal until 36.5sec when the stack was hit by a lightning strike followed by a second strike at 52sec. The crew noted unexpected motion in the vehicle following the discharge and severe effects were observed on the displays.

In the first discharge on the CSM the three fuel cells were disconnected, putting power on to two of the three entry batteries in the Command Module, a supply normally reserved exclusively for the end of the mission when the Service Module separated. The crew observed how the entire display console lit up with warning lights, audible alert tones and flashing indicators. The lower voltage took the signal conditioning equipment off line and that severed the telemetry link to the ground.

At the second discharge the CSM inertial guidance platform was lost but eight seconds later the ground got telemetry back and the crew was asked to switch to the secondary conditioning equipment so that they could diagnose the situation. That was activated at 1min 38sec into flight and the crew were instructed to reset all three fuel cells, numbers 1 and 2 coming back on line at 2min 21sec followed by fuel cell 3 at 2min 51sec. As the main bus voltage increased to 30vdc all the electrical systems came back on line.

Throughout the diagnostic and crew activity to restore order, the Saturn V, which was responding normally and was completely unaffected by the discharges affecting the CSM, was going through the final stages of the S-IC propulsive phase, the centre F-1 engine being shut down at 2min 15sec followed by the guidance platform arresting the tilt at 2min 38sec and shutdown of the outer four engines at 2min 42sec. In full analysis after the flight, meteorologists concluded that 'the lightning was triggered by the presence of the effective electrical conduction path created by the space vehicle and its exhaust plume in an electric field which would not [have] otherwise discharged'.

The event brought significant changes to the weather criteria for launch, most notably that 'no launch [will take place] when flight will go through cumulonimbus [thunderstorm] cloud formation…no launch if flight will be within 5 miles [8km] of thunderstorm clouds or 3 miles [4.8km] of associated anvil'. This, and a range of very specific criteria for providing approval for launch which set in motion a series of constraints which have come down through the decades to this day.

But apart from the way in which the event was handled, it vindicated the sometimes insistent approach taken by Wernher von Braun at the very beginning of the human space fight programme when he persistently resisted attempts by the Mercury 7 astronauts that they, and not an automated computer, should take responsibility for controlling the performance of the launch vehicle. By separating the control functions of launcher and spacecraft, the concept of the Instrument Unit (IU) emerged in which a separate bank of computers and guidance systems would be responsible for the management of the flight into orbit. Had that design philosophy not been installed in every programme from the beginning, the flight of Apollo 12 would have been aborted less than

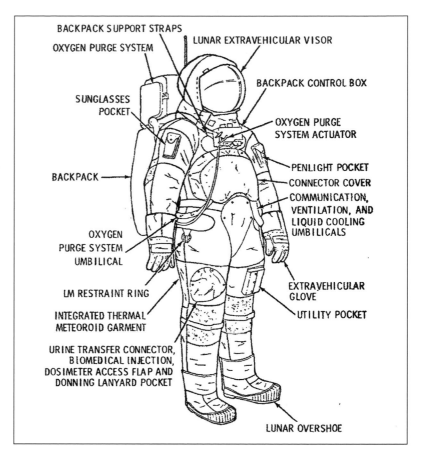

a minute after launch because the control of the entire vehicle would have been under the authority of that part of the stack that triggered the discharge.

ABOVE The Extravehicular Mobility Unit (EMU) for the H-series missions was modified from that employed for previous flights, with changes to the visor and to the arrangement of pockets. During this period, in workshops at contractor and NASA facilities an improved A7L-B suit will be made available for the J-series flights. *(NASA)*

LEFT Displaying the general configuration of the side-by-side seating, Dick Gordon tidies up inside the Command Module trainer and familiarises himself with the camera equipment. *(NASA)*

Despite the electrical discharges affecting only the Apollo spacecraft there were some concerns about potential electrical damage to the Lunar Module but there was no way of knowing that before departing Earth orbit so, with no obvious signs of damage, the mission proceeded on the basis that a determination would be made at an unscheduled checkout on the way to the Moon. However, additional checks were made of the CSM during the period between reaching orbit and firing up the S-IVB for the second time. The platform, which had been lost during the strikes, was aligned followed by two realignments to check for drift in the gyroscopes. Because there was little dark field and no bright stars in the field of view it was difficult to get stellar alignments but stars Rigel and Sirius were used and auto-alignments were performed satisfactorily.

Re-ignition of the S-IVB occurred at the nominal time of 2hr 47min 23sec for a 5min 41sec burn, accelerating the reduced stack to a velocity of 35,427ft/sec (10,798m/sec), 221ml (355km) above Earth. Procedures were as planned and without incident and the CSM separated from the SLA at 3hr 18min 05sec,

turning around and docking with Intrepid 8min 48sec later. At transposition and docking, CSM-108 weighed 63,535lb (28,819kg) and LM-6 weighed 33,584lb (15,234kg), a total mass of 97,119lb (44,063kg). Final separation occurred at 4hr 13min before the Auxiliary Propulsion System on the S-IVB stage fired to place it on a slingshot trajectory with the Moon.

The S-IVB would make its closest approach to the Moon at 85hr 48min, passing within 3,547ml (5,707km) of the lunar surface and receiving a boost of 0.34ml/sec (0.54km/sec) from that encounter which placed it into a highly elliptical orbit where it would continue to orbit the Earth rather than enter a heliocentric path as planned. This was the last S-IVB to be targeted for solar orbit. The remaining five S-IVB stages sent toward the Moon would be directed to impact the surface, providing shock waves for the seismometers to measure the interior structure of the Moon itself.

With a wide spectrum of hardware available for creating shock waves for detection by seismometers left on the lunar surface, scientists would receive data instrumental in calculating what the Moon was like

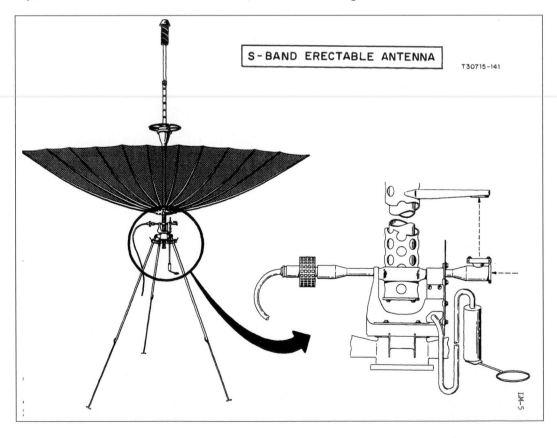

S-BAND ERECTABLE ANTENNA

T30715-141

RIGHT Not deployed on Apollo 11, the Erectable S-Band Antenna would be used to make a direct connection between the colour TV camera on the surface and the tracking stations on Earth, generally improving communications. *(NASA)*

deep beneath the surface, whether it was heterogeneous or homogeneous, whether it had a thick or a thin crust and what the upper and lower mantle was like. From the S-IVB to the Ascent Stage of the Lunar Module, expended hardware could be used for a secondary purpose even in its destructive impact with the lunar surface. As will be seen later, even mortars firing bombs and 'thumper' devices creating shock waves across just the thin outer layers of the surface would help build a coherent picture of the Moon.

At 7hr 20min into the mission, Conrad slipped through the docking tunnel into Intrepid for a checkout of the LM to make sure the lightning strike during launch had not affected Intrepid. All was well and little more than an hour later both he and Al Bean were back in the Command Module. But even as the crew got ready to set up the passive thermal control mode, technicians at the Manned Spacecraft Center noticed a 1amp overload on what they expected to see on the electrical supply, asking Conrad to go back into Intrepid and see if a light had been left on. At 10hr 35min the crew moved back to Intrepid and confirmed that the light did not seem to extinguish via a microswitch in the overhead hatch – so they pulled that circuit breaker and the 1amp disappeared.

The TLI burn placed Apollo 12 on a free-return path which would carry the spacecraft to a pericynthion altitude of 541ml (870km) around the far side of the Moon. But this would not enable the spacecraft to reach the designated landing site, for which it was necessary to transfer to a non-free return path. That manoeuvre was conducted in the knowledge that Intrepid's DPS engine could put the flight path back on to a free return type should the SPS engine become inoperable. In the first such hybrid transfer of the Apollo programme, the SPS engine conducted a 61.8ft/sec (18.8m/sec) burn at 30hr 52min 44sec to place the docked spacecraft on a far side pericynthion of 69ml (111km).

The mood aboard Apollo 12 was more relaxed than it had been on the preceding flight, largely due to the ebullient and frequently mischievous attitude of Pete Conrad, who ran his ship by consensus rather than command. But a lot of it was due to the fact

that the landing procedures had been tested, found successful and were safe; tension and expectation of possible failure had been removed by the flight of Apollo 11 and the fact that the engineering solutions had been verified and that Apollo was now shifting to an operational phase helped relax the crew.

But there were also the significant changes brought about by the more commodious living conditions aboard the Apollo spacecraft compared to the two-man spacecraft that Conrad and Gordon had experienced:

CONRAD: 'We are trying all these things we didn't have in Gemini, like toothpaste and shaving – we are really having a ball up here!'

CAPCOM: 'Roger. All dressed up and no place to go.'

CONRAD: 'Oh, we're going someplace. We can see it getting bigger all the time.'

And so it went on, as Apollo 12 repeated the translunar flight routines of the preceding two Apollo missions, conducting the two Lunar Orbit Insertion burns to place the two spacecraft in an orbit of 76.3ml x 62.9ml (122.7km x 101.2km).

ABOVE Carried in a bag from the side of the Lunar Module, the antenna was designed to be deployed only after the crew reached the Moon. *(NASA)*

RIGHT The antenna is
here being extended
before deployment
like an inverted
umbrella. It was felt
that its use on Apollo
11 would take up too
much time. Simulated
deployments such
as that shown here
helped mission
planners construct
timelines for surface
tasks, each of which
was assigned a
specific slot and task
duration. (NASA)

BELOW Fully open,
the antenna would be
stabilised and aligned
with Earth. (NASA)

Conrad and Bean entered the Lunar Module
to conduct a clean-up and stowage routine at
the same time as landmark tracking on a crater
in the vicinity of Fra Mauro, a site scheduled
for Apollo 13, before moving back to Yankee
Clipper less than two hours later. It was time
for a rest period so that the crew would be
refreshed for the landing ahead. Unlike Apollo
11, from the outset the first EVA was to be
conducted before the initial sleep period on the
surface and the need for a pre-conditioning rest
was vital; at more than 20 hours, the day would
be long and packed with activity.

The plan for executing a pin-point landing
began with undocking. Instead of a hard
undock that caused some orbital displacement
on Apollo 11, the two spacecraft conducted a
soft undock with physical separation involving
Yankee Clipper's RCS thrusters rather than
those on Intrepid, again conserving the known
trajectory of the LM. Neither did Intrepid conduct
a yaw manoeuvre or active station-keeping,
leaving all that to Yankee Clipper. And because
the landing point designator was to be used to
conduct a manual re-designation of the target,

the crew calibrated the platform by sighting on a star for which the descent trajectory was designed. And after DOI, the residuals were not to be nulled out but voiced down to the ground so that the errors could be factored in to the state vector update to more accurately place the descent path on the anticipated track.

Also, because of the more westerly landing site, the additional time around the near side of the Moon allowed Mission Control to uplink a new state vector based on the immediately preceding orbit which, coupled to the uplinked residuals factored in, provided a much greater degree of accuracy with no manoeuvring or propulsive activity to shift the actual flight path. In addition, unlike Apollo 11, where the crew were initially lying face down toward the Moon for landmark tracking, they began PDI face-up, eliminating the need for a 180° yaw manoeuvre to gain the correct orientation for High Gate.

Learning from Apollo 11, where structural blockage of the steerable antenna on the Lunar Module caused it to lose lock during some of the descent, Apollo 12 had corrected blockage plots so as to sustain steering lock with Earth. Not that coverage analysis had been lacking; whereas Apollo 10 had one volume of tracking orientation, Apollo 11 had five volumes each dedicated to a specific phase of the mission.

Mission rules required an abort in the descent phase if voice and high-bit-rate telemetry was not maintained and it was known that low signal levels could not connect to an omnidirectional antenna on the LM when transmitted to an 85ft (25.9m) diameter antenna at the Deep Space Network. As a contingency it was determined that if communication through the LM steerable antenna was blocked, satisfactory signals could flow from an omni antenna to a 210ft (64m) antenna on Earth and the landing was delayed by one revolution to allow for alignment with just such a station as the Earth rotated.

Powered Descent Initiation began with ignition of the DPS engine at 110hr 20min 38sec, followed by throttle-up at 27sec into the burn. Landing site corrections were put into the computer at 1min 25sec and again 24sec later, with landing radar altitude lock-on at 3min 22sec and a height of 41,438ft (12,630m), followed by velocity lock-on 4sec after that at a height of 40,100ft (12,222m).

The initial difference between the altitude measured by the landing radar and the onboard prediction of 1,700ft (518m) was resolved to 400ft (122m) within 30 seconds as the guidance computer took over to merge the discrepancies. During the early phase of powered descent the LM initiated a roll angle of 4° to steer out an orbital plane difference which was carrying the spacecraft 5.7ml (9.1km) north of the target, a ΔV of 78ft/sec (23.7m/sec) which was completed at five minutes into the burn.

All the way down to High Gate, at 6,989ft (2,130m) above the surface and 170ft/sec (52m/sec), the dispersions were nulled to close to zero. On automatic pitch-over for visual acquisition of the target, the computer generated LPD elevation angles and the crew reported almost immediately that they had the Snowman configuration of craters right ahead. The landing phase was performed manually following a P66 into the computer at a height of 368ft (112m) and a descent rate of 8.8ft/sec (2.68m/sec) with descent almost vertical from 50ft (15.2m).

The piloting skills of Pete Conrad came into focus when he commanded seven

re-designations of the landing site, manually
controlled with the rotational hand controller.
The total effect was to move the automatic
target point 718ft (219m) to the right and
361ft (110m) downrange of the initial target,
with the LM travelling a total of 1,500ft (457m)
downrange, about 400ft (122m) less than the
automatic targeting. But the ebullient personality
of Pete Conrad shone through, indicative of a
new, more confident and assertive assault on
the Moon, infectious to Alan Bean talking his
commander down to NASA's second landing on
the lunar surface:

BEAN: '40 coming down at 2. Looking good,
watch the dust. 31, 32, 30 feet. Coming down
at 2. Pete, you got plenty of gas, plenty of gas
babe. Stay in there.'

CAPCOM: '30 seconds.'

BEAN: '18 feet coming down at 2. He's got it
made. Come on in there…contact light!'

CAPCOM: 'Roger, copy contact.'

Contact occurred at 110hr 32min 36sec, some
11min 58sec after ignition of the DPS but one
second after engine cut-off based on the contact
light indication. As read out on displays in the
Mission Operations Control Room, the propellant
tank quantity gauge indicated by fuel tank No 2
read 5.6% remaining, equating to 1min 53sec

of engine time left. Post-flight analysis indicated
that the mean propellant readings discounted for
sloshing effects should have resulted in a low-
level reading some 25 seconds later with engine
shutdown 35 seconds after that, indicating
1min 16.5sec remaining. Because the indication
was received early the actual time remaining to
propellant depletion was 1min 43sec. Of 18,429lb
(8,359kg) of propellant loaded in the Descent
Stage, 1,175lb (533kg) remained versus 667lb
(302kg) predicted pre-flight.

During the computer-controlled portion
of the approach phase Conrad executed a
2° right correction which was made so as to
manoeuvre Intrepid away from the centre of the
Surveyor crater but the pre-flight targeted spot
at the 4 o'clock position of that crater seemed
too blocky and a 2 o'clock position was more
suitable. This degree of precision relocation was
something quite different to that experienced
by Neil Armstrong and a steeper descent to the
surface was made. The crew first noticed dust
at a height of 175ft (53m), which gradually built
up in radial fans away from the downdraught of
the LM's descent engine. The last 50ft (15.2m)
was conducted totally in the blind, based on
visual designation of the exact spot chosen
when last observed from Intrepid.

During the final descent phase the LM
came within 357ft (109m) of the Surveyor III
spacecraft, sitting on the south-east inside slope
of a crater 656ft (200m) in diameter. Intrepid
touched down on a flat surface, 508ft (155m)

from Surveyor III on the north-west rim of the crater – the 2 o'clock position. Material blown laterally by the plume of the descent engine had been deposited on elements of the spacecraft brought back to Earth, providing an additional analogue on the traverse range of such particulates and on the amount of scouring that could be expected from rocket plumes under the lunar environment. But that was ahead of them as the crew made an initial observation of their surroundings and summarised the scene, recorded in post-flight debriefings:

'At first glance out of the spacecraft window, there was absolutely no distinguishable colour difference. About the only difference noticed was in looking cross-Sun versus looking down-Sun. There were no immediately apparent white rim craters near us. Most of the craters observed from the LM window did not have any particular elongation. The craters seemed to be the same texture as the area surrounding them. All the material looked the same until we were very close to the individual rocks.'

But in the minutes after landing, the two astronauts were not at all sure precisely where they were, the low Sun angle providing no shadows to help delineate the subtle undulation and shallow craters that would provide a positive reference to the Snowman. It soon became apparent that they were in fact right on target when Gordon called from Yankee Clipper passing overhead and reported, 'I have Snowman. And I believe I have the (LM) on the northwest side of the Surveyor crater.'

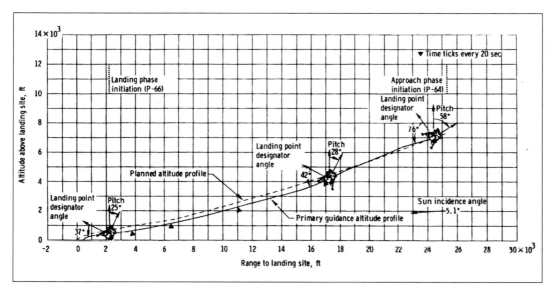

LEFT The geometry of the actual descent trajectory down to the Moon versus the planned trajectory. The landing designator angle is shown compared to the pitch of the Lunar Module out of vertical. *(NASA)*

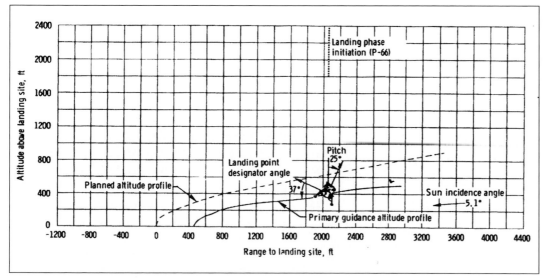

LEFT Showing events closer to the surface, this chart takes the LM closer under P66 with the divergent planned altitude profile versus the profile from the Primary Guidance, Navigation and Control System (PGNCS). *(NASA)*

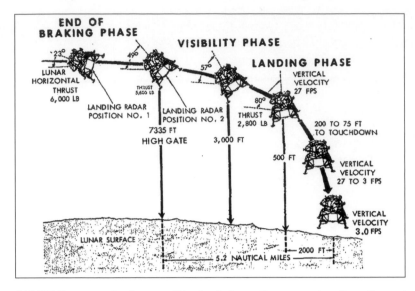

ABOVE For general reference, this chart shows the parameters from the end of the braking phase, on through the beginning of High Gate and into the phase where the site becomes visible to the Commander looking out the left window while the Lunar Module Pilot keeps his head down, reporting data from the displays. *(NASA)*

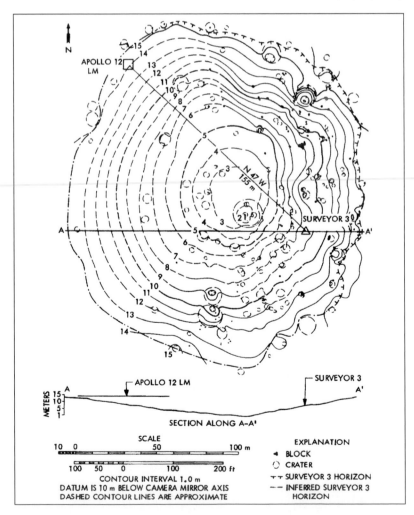

The somewhat underplayed encounter between humans and a lunar robot was seminal in teaching scientists and engineers how valuable is the interaction between manned and robotic research. NASA was delighted that it had pulled off another spectacular success and all nervousness evaporated when it was realised that Intrepid was exactly where it was supposed to be – perhaps even a little closer than planned to the Surveyor III spacecraft.

Immediately after landing the propellant tanks were vented and the vehicle made safe, with a 'go' for stay at T1 and T2, prior to configuring Intrepid for a planned surface stay of about 31.5 hours with two EVAs of 3.5 hours each separated by approximately nine hours of eating and sleep before leaving the surface to re-join Yankee Clipper. The EVA timelines were flexible and any extension was contingent on the quantity of consumables remaining in the PLSS backpacks, which was dependent on the energy expended by the astronauts on their activities, a level rated in British Thermal Units (BTUs) or kilojoules/hr. There was provision for a 30-minute extension should energy levels prove lower than estimates. For example, Conrad was expected to expend 4,081BTU (4,301kj/hr) on the first EVA and 4,235BTU (4,464kj/hr) on the second.

The overall priority for getting the ALSEP instruments deployed placed it on the first Moonwalk after contingency and bulk sample collections, with the secondary objective of reaching Surveyor III as part of a geological sample collection traverse on the second EVA. Although considered the first operational mission, H-1 had a significant increase in the duration of both the surface stay and the duration of the EVAs, the latter providing up to 16man/hr compared to 5man/hr on Apollo 11.

While public attention began when the astronauts left Intrepid, a significant amount of preparation preceded that and, as with Apollo 11, it took longer than expected to complete all the configurations required to prepare the

LEFT A contour map of the Surveyor Crater separating the landing spot for Apollo 12 and the resting place for the Surveyor III spacecraft on the inside, south-eastern wall. *(NASA)*

LM and each crewmember. Hallmark of a true professional, in technical debriefings after the flight Conrad was self-critical about his impatience to get outside and admitted leaving the rigid sequence of procedures in an attempt to reduce the time taken to complete all the required processes, while congratulating those whose job it had been to prepare the checklists and cue-cards.

The first EVA officially began when the hatch was opened at 115hr 10min 35sec, some 90 minutes later than planned due to an extended description of the surrounding area which had taken place so as to definitively tie down precisely where Intrepid was on the surface and how it sat in relation to the Surveyor spacecraft. Diminutive in stature, with both hands on the ladder, Pete Conrad hopped down from the lowest rung and landed in the footpad, 22 minutes after the start of the EVA. But there were no symbolic words to match those from Neil Armstrong:

CONRAD: 'Whoopee! Man, that may have been a small one for Neil but that's a long one for me! I'm going to step off the pad. Mark. Off the… ooo, is that soft and squeezy. Hey, that's neat. I don't sink in too far. I'll try a little – boy, that Sun's bright. That's just like somebody – shining a spotlight on your hand. Well, I can walk pretty well, Al, but I've got to take it easy and watch what I'm doing. Boy, you'll never believe it. Guess what I see sitting on the side of the crater. The old Surveyor; yes sir. Does that look neat! It can't be any further than 600 feet [200m] from here. How about that?'

First order of business was to collect a contingency sample in case of an early termination of the Moonwalk. This was followed by deployment of the MESA table and the equipment transfer bag (ETB) so that Conrad could use the lunar equipment conveyor to move the sample together with the PLSS batteries and LiOH canisters, which had been carried in Quad IV, up to Alan Bean who stowed them inside the LM for the second EVA.

The TV camera used to show Conrad descending the ladder had been carried on Apollo 10 in the Command Module and was the first use of colour TV on the Moon. The electrical connection between it and the existing wiring

loom in Intrepid was made via an adapter so as to simplify the upgrade required to carry the colour signal into the LM and across to Earth via the S-band antenna. Changes to the camera included painting it white for thermal reasons and replacing coated metal gears for plastic ones.

As Bean began to move outside on to the porch some 40 minutes after Conrad his backpack tore part of the insulation on the outside of the hatch. A little ahead of the timeline, the two busied themselves first by setting up the erectable S-Band antenna – a large umbrella-like 'dish' folded into a cylindrical case attached to the exterior of the Descent Stage on Quad I adjacent to the ladder. It was through this that communications would be transmitted to Earth. Set up on a tripod, it took two men to begin unfolding and erecting it starting about 43 minutes into the EVA and within 15 minutes it was erected.

This 10ft (3m) diameter antenna had been

BELOW Pete Conrad descends to the surface of the Moon, photographed by Bean inside the Lunar Module, a rare shot of an astronaut going down the ladder as viewed from inside the hatch. *(NASA)*

carried on Apollo 11 but at the time it was
estimated that it would take at least 19 minutes
to set it up and that was just too long for the
crowded schedule on that first Moonwalk.
The need for this antenna was driven by
the wideband frequency modulation, which
appreciably reduced background noise while
marginally increasing signal strength. But the
difficulty in receiving these downlinks from the
Lunar Module pushed engineers to develop
the erectable antenna which was capable of
handling a stronger TV signal. The longer time
required to set up the erectable antenna was
the reason why the degraded signal from Apollo
11 was deemed sufficient for that first landing
and why the public relations people at NASA
looked forward to much better TV pictures from
the Moon and in colour on this second landing.

An early activity involved moving the TV
camera from the MESA to the erectable tripod
where it was supposed to look back at the
Lunar Module and capture the raising of the
flag. In doing so, that orientation made it difficult
not to point the camera up-Sun when placing
it on the tripod and, about 50 minutes into the
EVA, Al Bean did just that and despite attempts
to get the picture back the only image that
came through was one in which a large band of
white appeared at the top while the rest of the
frame was dark.

One of the reasons why the TV camera was
inadvertently pointed at the Sun was due to the
pre-touchdown yaw manoeuvre which placed the
LM directly down-Sun. But the precise reason
was uncertain until after the mission. For the rest
of the surface activity there would be no TV, a
public relations disaster. NASA was fully aware
that the media would quickly lose interest in
repeated Moon landings unless stimulated by
pictures and colour views of the two Moonwalks.
Although, in reality much of the crew activity on
the surface would be out of camera range.

RIGHT The Erectable S-Band Antenna deployed
adjacent to the Lunar Module ready to broadcast
colour pictures to Earth. (NASA)

If a pin-point landing to a meaningful geologic site had been the prime goal of the Apollo 12 mission, another prime objective had been to deploy the Apollo Lunar Surface Experiments Package (ALSEP). ALSEP was the first significant shift toward aligning the lunar landings with a more enduring value.

Despite the political redirection of Apollo as a political goal for the Cold War there was abiding recognition that it never would be sufficient just to race to the Moon and leave. There was a fundamental desire to understand the Earth–Moon system and to do that, human access to the lunar surface was an implicit part of NASA's original mandate. The scientific fraternity wanted it that way and ALSEP was the logical start to the process by leaving behind a suite of instruments that could continue to monitor the lunar environment long after the visitors returned to Earth.

BELOW The Westinghouse colour TV camera which was set up on the Moon to show viewers activity around the Lunar Module Intrepid and to be repositioned to show the deployment of the ALSEP experiments. *(NASA)*

The formal beginning of the ALSEP programme was on 31 March 1963, in a decision reached between NASA headquarters, the Jet Propulsion Laboratory and the Goddard Space Flight Center in which a range of experiments would be selected for deployment by the crew. By this date the decision had been made to use a single Saturn V per mission and to employ a separate lander, allowing multiple visits and a range of hardware. By the end of 1963 general agreement had been reached about the type of experiments to produce, based on minimum weight, simplicity of design and ease of deployment.

High priority was given to active and passive seismic measurements of the Moon's interior, to measurements of the Moon's magnetic field, to observations of the general environment of the Moon within the presence of the solar wind, with an added desire to understand the bearing strength of the surface and to the density of the outer layers. The preliminary selection had been made by the end of 1963 and at a meeting of the National Academy of Sciences' Space Science Board in July 1965 a core framework was agreed upon which the ALSEP programme would proceed.

Major decisions were made in 1966: on 14 February the first published list of experiments and their assigned missions was presented to the Manned Spacecraft Center; on 16 March the Bendix Corporation was selected to carry out all the design, construction and testing of the various packages; and on 6 November the contract was approved and a delivery date of July 1967 set for the first suite of what was considered a rolling production of tailored experiment packages for successive landings.

Because Apollo 11 (the only G mission) was considered an engineering test, the culmination of a development phase which only incidentally also provided the first crewed landing on the Moon, as related earlier the much more simplified EASEP package was carried and deployed. It did not have the power source or the integrated data transmission station which would characterise the more ambitious and productive ALSEP arrays, the decision having been made that one of these would be carried on every Moon landing.

Common to all ALSEP arrays was the Central Station (CS), the Radioisotope Thermoelectric Generator (RTG) and the cables connecting the two, in addition to a network of data cables connecting the CS to each science instrument laid out across the lunar surface. The complete ALSEP array was contained in two separate sub-packages inside Quad II, which formed the Scientific Equipment (SEQ) bay at the left rear of the Descent Stage when looking forward, diametrically opposite the Modularised Equipment Stowage Assembly in Quad IV.

The CS was to be set up on the lunar surface as the central hub of the array and would be utilised for sending data to Earth and receiving commands for the various experiments. With a weight of 55lb (25kg) the CS was not easy to deploy as it basically consisted of a concise package of electronic equipment inside a flimsy box-shaped structure serving as a thermal shield. With an internal volume of 34,800cm³ (2,123in³) it contained a 74W power control unit with an output of 53–63W at various voltage levels and a data subsystem transmitting at 1W on 2275.5mHz, 2276.5mHz or 2278.5mHz at 530, 1,060 or 10,600bits/sec.

Commands from Earth were accepted at 2119mHz. Communication was via an axial-helical antenna with a length of 23in (58cm) and a diameter of 1.5in (3.8cm). Average data rate was 16.55bits/sec and it was capable of decoding 100 discrete commands for users. Carried on both sub-packages, elements of the CS comprised separate base and side curtain panels topped off with a Sunshield.

Power for the ALSEP was provided by the Systems for Nuclear Auxiliary Power (SNAP) programme of the US Atomic Energy Commission, which had been set up in 1946 to handle the peaceful development and application of nuclear power. Under AEC nomenclature, even-numbered projects were for full-scale nuclear reactors while odd-numbered systems were RTG designs. The RTG was based on the principle that a nuclear power source would operate as a fuel capsule immersed within a series of concentric rings of thermocouples which would produce electricity through the radioactive decay of Plutonium-238 dioxide, which has a half-life of 87.7 years.

Known as SNAP-27, when fuelled the RTG employed for ALSEP weighed about 44lb (20kg) and consisted of a cylindrical generator fabricated from beryllium, approximately 18in (45.7cm) high with a diameter of 16in (40.6cm). The Pu-238 pellet was 16.5in (41.9cm) in length with a diameter of 2.5in (6.3cm). While the cask, thermal-rejection fins and base upon which it would stand on the lunar surface were carried on one of the two ALSEP pallets in Quad II, the fuel itself was carried in a graphite cask attached by a titanium structure to the left side of the SEQ bay, cantilevered so that it could be rotated from a vertical to a horizontal position, removed and inserted into the RTG.

The original structure of the ALSEP programme was to provide four sets for the H-series flights followed by a determination as to requirements for the five planned J-series extended missions. The cancellations and adjustments would provide for only three H-series and three J-series but all were successful and provided many years of solid data upon which to build a more comprehensive picture of the lunar environment from science stations down on the surface.

BELOW A diagram from the astronaut training manual shows the various sections of the TV tripod and its adjustments. *(NASA)*

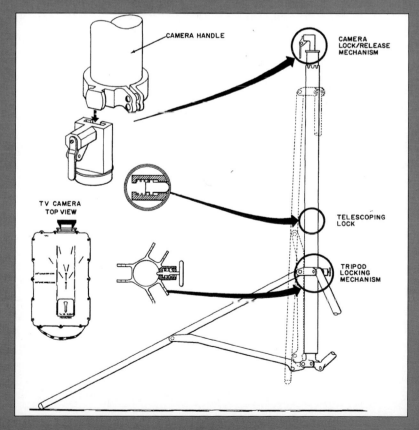

CAMERA HANDLE

CAMERA LOCK/RELEASE MECHANISM

TV CAMERA TOP VIEW

TELESCOPING LOCK

TRIPOD LOCKING MECHANISM

RIGHT Bean inadvertently pointed the camera at the Sun which damaged the potassium chloride and burned out the vidicon tube which prevented any further TV transmissions from the surface, losing a big audience on Earth. *(NASA)*

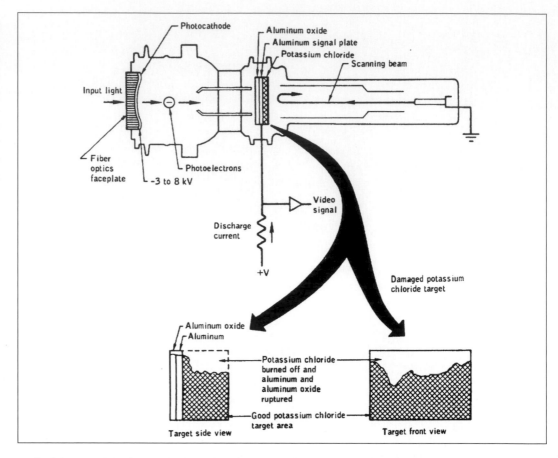

And then, at 50 minutes the two set about removing the ALSEP sub-packages which were deployed to the surface at 116hr. It was Bean's job to extract the Pu-238 fuel from its graphite cask and insert it into the RTG, which Conrad removed from the Scientific Equipment Bay around the back of the LM. But Bean had great difficulty removing the plutonium pellet and only by tapping on it several times could he prise it loose and insert it in the finned thermal generator. But the delay in getting it out was noticeable, heat from the radioactive pellet flowing through the astronaut's gloves.

The two sub-packages were connected by a barbell, an arrangement known to weightlifters as they raise two opposing weights connected by a bar. While comfortably within the lifting capacity of a suited Moonwalker, Bean found the loping motion necessary for traversing the surface made it difficult to maintain a grip on the bar, suggesting several improvements for future flights. It was down to the astronauts to choose a suitable deployment site: not on an uneven surface, not on a slope or within proximity of potential shadow from rocks and at least 300ft

(100m) from the LM so that the blast of lift-off would not throw quantities of dust across the array; all the instruments had been carefully designed for thermal properties anticipated on a clear and exposed surface. The astronauts chose a position 425ft (129m) from the LM and set up the power cable to the Central Station before connecting it to the RTG.

The first science instrument to be set up was the Suprathermal Ion Detector Experiment (SIDE) in association with a Cold Cathode Ion Gauge (CCIG) which together were designed to study charged particles in the lunar environment and to provide a reading on the electrical potential of the lunar surface. The SIDE also measured neutral particles and provided an opportunity to determine the chemistry and radioactivity of the Moon but the readings would also allow measurement of the solar wind itself and of the effect of the Earth's magnetosphere as it swept across the surface when the Moon was on the opposite side to the Earth with respect to the Sun, a time of Full Moon. The instrument weighed 19.6lb (8.89kg) and required an operating power of 60W.

Next came the Passive Seismic Experiment (PSE), difficult to deploy due to the mounting stool of this drum-shaped device which prevented a proper contact with the lunar surface – essential for an instrument designed to monitor shock waves through the outer layers of the Moon. Renowned geologists and Earth scientists Gary Latham, Maurice Ewing and Frank Press were behind the PSE and designed the instrument which consisted of three separate sections: the tri-axis detectors consisting of capacitance sensors, two for the horizontal axes and one vertical axis; a levelling stool for fine adjustment by the astronaut to within +/-5° with final levelling from control motors to within 3arc-sec; and a thermal shield like that placed directly over the instrument.

The Solar Wing Spectrometer (SWS) was set up on the lunar surface to detect and measure electrically charged particles which are prevented from reaching the surface of the Earth because of the planet's magnetic field, although they can reach the upper atmosphere in the familiar form of auroras. The SWS comprised seven modified Faraday cups, one to the vertical axis with six surrounding the vertical where the angle between any two adjacent cups was approximately 60°. Because each cup measured the current produced by the particle flux the measurements would be identical if it is equal in each direction. If not, the amount of current could be measured for each cup to determine the variation in particle flow according to direction.

Supplementary to the SWS, the Solar Wind Composition Experiment which had been deployed by the Apollo 11 crew and returned to Earth for analysis was also carried on Apollo 12 but instead of two hours' exposure on this second landing it was retrieved only on the second EVA after an exposure of around 17 hours.

The Lunar Surface Magnetometer (LSM) was the last ALSEP instrument to be deployed, equipment which could tell much about the very nature of the Moon itself. Designed to measure the magnetic field at the surface, the LSM consisted of a tri-axis flux gate magnetometer with three sensor arms aligned in orthogonal axes and mounted to a central structure. Inside this structure are an electronics bay and an electro-mechanical gimbal/flip unit which allowed the sensor to be pointed in any desired

direction for survey or calibration modes. The astronaut used a shadowgraph to align it to within +/-3° east or west and to within the same margin of the vertical with the use of a bubble level. The LSM weighed 17.5lb (7.9kg) and when deployed with the arms extended 40in (101.6cm) high with 60in (152cm) between sensor heads at the tip of each boom.

The LSM could be used in either site survey or scientific mode. The LSM would be employed initially for site survey which identified any localised magnetic influences so that the measurements could be subtracted from the scientific readings. It was already known that the Moon did not have a measurable magnetic field but that did not discount the possibility of a localised field. The scientific mode activated all three sensors for simultaneous readings and for providing outputs proportional to the incidence of the field components. Measurements would be taken three times each second, which was a rate faster than any localised field was likely to change.

Inert up to this point, the next job was for the two Moonwalkers to begin to set up the Central Station and install the antenna, align it

ABOVE The Solar Wind Composition experiment involved rolling out an aluminium sheet which allowed solar wind particles to embed themselves in the foil and, when returned to Earth, provided scientists with a detailed analysis of their composition. With the exception of Apollo 17, every mission carried one and that for Apollo 16 used platinum foil. (NASA)

with Earth, with Conrad activating the shorting switch, turning on the transmitter and starting back toward the Lunar Module. Bean's job was to photograph the ALSEP site and the equipment and conduct additional sample collection. They had already been given a 30-minute extension to the nominal 3.5hr EVA and used this time to move across to the rim of a crater informally called 'Shelf', a 1,000ft (300m) depression north-west of their landed position. But the real work on this first EVA had been to set out the ALSEP.

The experiment package transmitter was turned on from Earth, 1hr 9min after the fuel capsule had been placed in the RTG, and it settled down to deliver 73.69W at a consistent level. The first EVA ended at 119hr 06min 38sec. The experiments were activated sequentially between 118hr and 124hr mission elapsed time beginning when the astronauts were still out on the surface and ending when they were inside Intrepid, having completed their first Moonwalk in a duration of 3hr 56min 03sec.

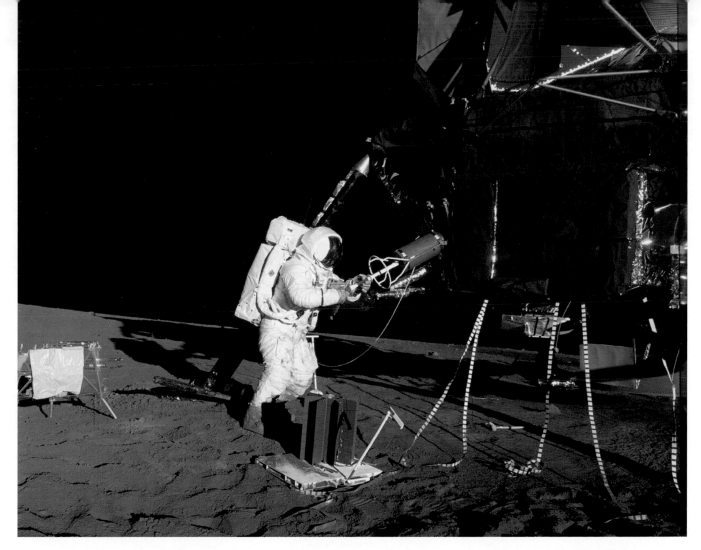

While the LM crew were preparing to recharge their PLSS backpacks from the environmental control system in Intrepid, Dick Gordon fired Yankee Clipper's SPS engine at 119hr 47min 12sec for 18 seconds to shift the orbital plane 4°, placing the Apollo spacecraft in a suitable orbit for ascent and rendezvous.

Margins built in to the EVA plan for consumables carried in the backpacks were adequate: Conrad had 42% oxygen, 44% cooling water and 34% battery power remaining and Bean's pack had greater reserves. Accordingly, metabolic rates for each crewmember had been less than expected, hence the lower drawdown on consumables, and the men were confident that they could have remained outside for at least an hour longer. But that was not possible due to the limited stay-time on the surface before lift-off and a sleep, together with a second EVA, which had to fit in to the relatively tight schedule for this mission.

As Armstrong and Aldrin had discovered on their flight, it was difficult to settle down after a Moonwalk and, while excitement was present,

even the professional approach and the debriefing kept them in long conversation with Mission Control about the surface, the materials they had sampled and the general layout of the site. All this was vital for scientists to absorb for the coming field traverse. The prime objective for this second EVA had always been a field geology traverse, the precise route for which was to be determined by the relative position of Intrepid to Surveyor. On the basis of the crew's report, mission planners prepared a traverse that would take them on a refined route, toward the end of which they would find themselves at the Surveyor spacecraft.

Conrad and Bean were two hours late, finally beginning their rest period at the end of a long day, at the beginning of which they had been awake since about 100hr 30min, to conduct activation of Intrepid, separation from Yankee Clipper, the landing and the first Moonwalk. Mission Control finally signed off at about 122hr 30min with the flight surgeon in Mission Control reporting that Conrad and Bean were soon

ABOVE Bean pulls the RTG fuel cask out of its mounting on the side of the Descent Stage in quad 2. *(NASA)*

Bean finds it difficult to extricate the plutonium core, reporting that he can feel its heat through his gloves. The finned RTG can be seen at right. *(NASA)*

sound asleep, some 22hr after their last wake-up call.

Little more than 6hr 15min later they were up and claiming to be 'hustling along' and that their rest had been 'short but sweet'. First, a meal and then a conversation with Houston over the upcoming EVA. But Grumman had been working some numbers and because Mission Control wanted the failed TV camera brought back – it would normally be left on the surface – the weight of rocks and samples returned should be reduced by 15lb (6.8kg). Not that the capability of the rocket motor would be compromised by such a small amount; rather that the combined effects of changes to the centre of gravity and the guidance equations for lift-off and ascent would change ever so slightly.

On completing their pre-EVA checklists, Conrad was confident on getting out the LM sometime before 132hr elapsed time, about an hour ahead of the Flight Plan despite a late start for EVA-1. The crew had made up time by completing several non-critical procedures and getting ahead of the pre-flight schedule in that they would compress a planned 15hr gap between the end of EVA-1 and the start of EVA-2 to less than 12.5hr, the official start of the Moonwalk being 131hr 32min 45sec. This day would be long, including the completion of a near-4hr EVA, preparations for ascent and lift-off plus rendezvous before rejoining Gordon in Yankee Clipper.

But it marked a seminal moment in the history of the Apollo programme, being the first EVA in which the sole objective was the geological exploration of a landing site and retrieval of parts from a spacecraft which had been sent to the lunar surface to answer crucial questions before humans followed. It was the beginning of a period of exploration that would last three years and see a total of five expeditions sent to the Moon for purposes of science while satisfying a deep-seated curiosity in humans to reach out and search new lands.

The procedures began with transfer of equipment from inside Intrepid to the surface, including the 70mm cameras, with Conrad removing tools from the MESA and stowing them on the Apollo Lunar Hand Tool Carrier (ALHTC), a folding structure made of aluminium with an unloaded weight of 9.3lb (4.2kg), a

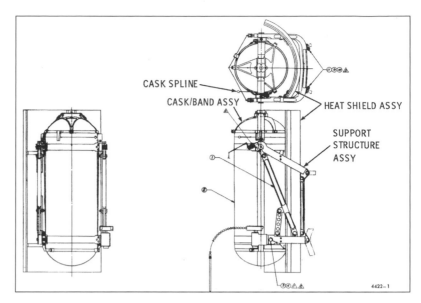

ABOVE A diagram of the RTG cask and associated support structure which allowed it to be pivoted downwards for extraction of the plutonium pellet. *(NASA)*

height of 26in (67cm), a length of the side with feet of 27.5in (70cm) and a width of 16in (41cm). It would carry small tongs, a short extension handle, 0.8in (2cm) diameter core

BELOW The top of the cask dome with restraints and locking cap. *(NASA)*

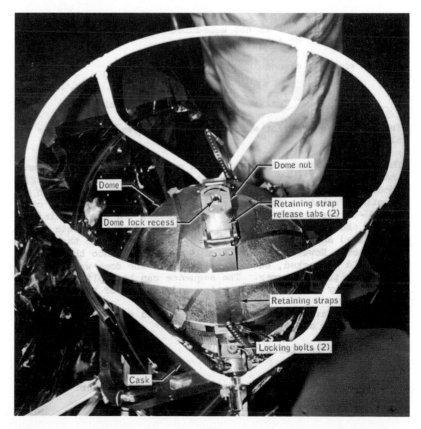

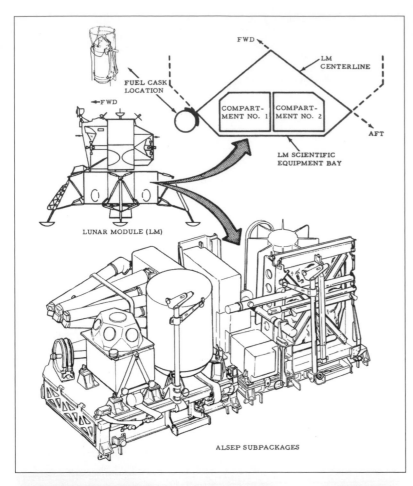

FWD

LM CENTERLINE

FUEL CASK LOCATION

FWD

COMPARTMENT NO. 1 | COMPARTMENT NO. 2

AFT

LM SCIENTIFIC EQUIPMENT BAY

LUNAR MODULE (LM)

ALSEP SUBPACKAGES

tubes and caps, round and flat rectangular documented sample bags and dispensers, a small non-adjustable scoop, lightweight hammer, brush-scriber-lens and a gnomon. A tote bag was provided made from a white woven cloth with a laminated Teflon covering.

Conrad was on the surface seven minutes after the official start of the EVA and Bean joined him ten minutes later. Moving past the ALSEP instruments, the seismometer picked up their loping walk as vibrations through the regolith. 'Pete, we're watching you down here on the seismic data. Looks as though you're really thundering right by it,' commented Mission Control. Their initial destination was Head Crater, west-south-west from Intrepid and about 150ft (50m) away. It was to the north-west of this crater that they had laid out the ALSEP and around on the north-west rim of this depression, which was only about 400ft (122m) in diameter, Conrad rolled a rock down its slope with his foot and then took a photograph of the track it made. And then both men stopped as Conrad rolled a larger rock and Mission Control noted the seismic vibrations while more pictures were taken to document the event.

From Pete Conrad: 'I've been concentrating Houston, as I came walking over here to Head Crater, to see if there is any possible change in either texture, slope, colour, anything you can think of, that would say to me that I was walking on a different surface than I was when I started. And I can't identify a thing yet, it all looks the same.' With a shovel attached to an extension handle, the crew then dug a small trench to obtain samples from a few inches beneath the surface but at one hour into the EVA they were hurried along, Mission Control

LEFT With ALSEP sub-packages in the foreground, the ALHTC can be seen to front left. *(NASA)*

ever watchful of the timeline.

On the way to Bench Crater, about 800ft (244m) south of the north-west rim of Head, Bean noticed some interesting samples: 'These rocks obviously came out of the crater, because they are scattered more uniformly around it. There's a bunch of them on the rim and there's not many far away. We probably ought to grab a big one of them.' But all the while time was against them and after surveying the area between Head and Bench, Bean quickly summarised: 'I noticed when I was looking at that rock back there up real close that it had been hit by meteorites so much I guess it had given it a rounded appearance, something like those in the hole except there's a couple over there like you say that don't look that way.'

Then it was time to head west-south-west toward Sharp Crater, for a ten-minute stop with core tubes and sample bags at this 30ft (10m) depression with a rim which Conrad noted was much softer than others in the vicinity. The two astronauts were now 1,300ft (400m)

from Intrepid, as far as they would get from the Lunar Module. At 1hr 50min into the EVA they packed up and headed east, passing the southern flank of Bench Crater on their way to Surveyor Crater and stopping halfway at Halo, a small depression which was difficult to identify as a crater in the long shadows of the rising Sun. And then they walked across to the south-western flank of Surveyor Crater, 650ft (200m) in diameter and the spacecraft that had been sitting there for 31 months.

Mission Control: 'OK, Pete. You're two hours and seven minutes into the EVA. And we show you leaving Halo at around 2:15 (EVA elapsed time). And now that's for a four hour EVA. We extended you 30 minutes for a total EVA of four hours. We'd like before you go on, to figure out a plan of attack on the Surveyor. Make sure… you remain from directly below the Surveyor as you move up to it.' There was some concern that the spacecraft could be unstable, sitting as it was at an angle of 12° and that was not unfounded. Post-flight inspection of the data

ABOVE Bean picks up the two ALSEP sub-packages and begins to walk them to their deployment location. *(NASA)*

together with extensive photography of the Surveyor III revealed that at some point since it landed it had slipped slightly, a motion probably induced by a minor seismic event or impact nearby.

The path to Surveyor was traversed from one slope across to the other and both astronauts tested the stability of the loose surface material to see that they could get back up out of the crater the same way they came – just in case they walked into trouble, perhaps a section of deep soil. Then they rested for a few minutes while looking across the crater's panorama and the precise position of the Surveyor spacecraft. When they finally reached it a number of tasks confronted them: remove and collect glass from the thermal boxes; photograph scoop marks on the lunar surface; cut and retrieve cables for return; and cut off the TV camera. The latter took the crew ten minutes to complete and place in Conrad's bag. Then it was time to return to Intrepid and button up.

Back at the LM, 3hr 10min into EVA-2, the crew stowed the solar wind collector, which had been out on the surface for 19hr, hauled a sample return container aboard the spacecraft and climbed back up the ladder. By the time the hatch had been closed at 135hr 22min for an EVA lasting 3hr 49min 15sec the two Moonwalkers had accumulated a total of 7hr 45min 18sec on two Moonwalks, more than three times the duration of the single EVA on Apollo 11. All this activity had been a magnificent start to lunar surface exploration and the crew were ahead of time, having corrected a somewhat extended preparation time for the first EVA.

Just as the crew got under way with their post-EVA checklist, Dick Gordon in Yankee Clipper was informed by Mission Control that observers on Earth had reported a transient event in the crater Alphonsus. Transient Lunar Phenomena (TLP) sightings had been reported at various times for several decades,

BELOW Bean lopes across the surface, leaving the Lunar Module behind as he transports the ALSEP array. (NASA)

characterised as brightening or sometimes an obscuration of certain crater floors as seen through the telescope. Nobody knew what they were but Apollo was responsive to these timely sightings and as the crater in question was under the CSM ground track it would be a good opportunity for the astronaut to get some photographs. Gordon failed to see anything but took pictures nevertheless.

Then Conrad and Bean hooked up to the Lunar Module's suit circuit, opened the front hatch a third time and dumped excess equipment on to the porch and down the ladder before closing up and re-pressurising the cabin. Before lift-off, there was time for a meal as both crewmembers actively described their EVA sessions to attentive listeners on Earth, discussing surface conditions and rock samples. But Gordon too had been busy. Equipped with a multi-spectral camera assembly he shot reels of film covering large areas of the lunar surface and obtained

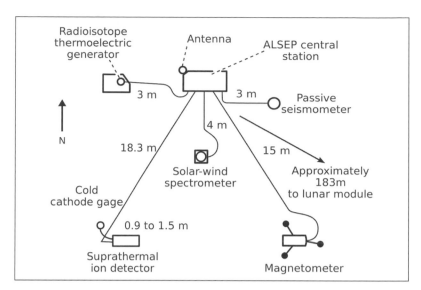

Hasselblad views of future candidate landing sites in the Fra Mauro and Descartes regions.

Fra Mauro was the tentative site for Apollo 13, a difficult area to get in to but one which would explore vast sheets of ejected material thrown out close to the beginning of the Moon's history when

ABOVE The Apollo 12 ALSEP instruments array as laid out on the lunar surface. *(NASA)*

LEFT The Lunar Surface Magnetometer which sought data on the Moon's magnetic environment, a body that does not have any polar dipole. *(NASA)*

ABOVE The Passive Seismic Experiment would record vibrations caused by internal activity, by impact of micrometeorites and by seismic movement caused by alignments with the Earth. *(NASA)*

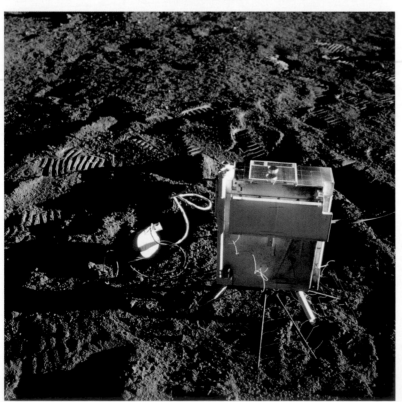

a massive crater 560ml (900km) across called a basin was excavated by a large object slamming into the surface. Called Mare Imbrium, the basin material extended far to the south and Fra Mauro was the type name given for this Imbrian ejecta blanket. So began a process of using one flight to reconnoitre another.

After 31hr 31min on the lunar surface, at 142hr 03min 48sec Intrepid lifted away to rendezvous with Yankee Clipper. Ten seconds before engine ignition Pete Conrad pushed the abort stage button which mechanically separated the Ascent Stage from the Descent Stage; five seconds later the Ascent Stage engine was armed and the PROCEED button was depressed, telling the computer's Luminary 116 programme to go ahead with ascent activity. At the time of ignition Conrad pressed the ENGINE START button to back up the automatic sequence and the motor roared into life, albeit unheard by the crew. 'Lift-off and away we go' were the words from Intrepid as the Ascent Stage lifted cleanly away from the site in Oceanus Procellarum.

Ascent events followed closely those demonstrated by Apollo 11 when CSI, CDH and TPI manoeuvres were performed as described for that mission. The only minor anomaly had been a 1.2sec overburn by the APS engine caused by a late positioning of the ENGINE ARM switch inhibiting the automatic cut-off, a situation in which the residuals were nulled by the crew through the RCS thrusters to keep the Ascent Stage in the pre-selected path. With a near-nominal rendezvous sequence, docking was achieved at 145hr 36min 20sec and Conrad and Bean joined Gordon back in Yankee Clipper.

The procedures for de-orbiting Intrepid began with LM jettison at 147hr 59min 31sec followed by a CSM separation manoeuvre at 148hr 04min 31sec, a 5.4sec RCS burn for a velocity change of 1ft/sec (0.348m/sec). Intrepid's de-orbit burn occurred at 149hr 28min 15sec. This was designed to crash the stage

LEFT The Suprathermal Ion Detector Experiment (SIDE) was positioned alongside the Cold Cathode Ion Gauge (CCIG) and 60ft (18m) away from the Central Station. *(NASA)*

Data and communications between Earth and the ALSEP experiments went via the Central Station, seen here with its pole antenna but not yet with side panels to protect it from the Sun. *(NASA)*

back on to the Moon 5.7ml (9.1km) east-south-east of the landing site for triggering seismic shock waves.

The RCS thrusters fired at 149hr 28min 15sec in a 4-quad burn lasting 1min 22sec, decelerating the LM by 196ft/sec (60.4m/sec) for a surface impact 27min 01sec after ignition. But a 2-second, 5ft/sec (1.5m/sec) overburn brought it down to the surface 46ml (74km) further downrange, impact occurring at 149hr 55min 16sec. The Ascent Stage struck the surface with a velocity of 3,755mph (6,043kph), making a crater calculated to be 36ft (11m) long by 20ft (6m) wide and approximately 16.5ft (5m) deep.

Scheduled to start their sleep period at 150hr elapsed time, the crew were intensely busy trying to clean up the interior of Yankee Clipper from the dust brought in on the sample return containers and the equipment brought up from the surface. Three hours late in beginning their rest, for Conrad and Gordon that day had lasted more than 25 hours. After barely four hours' rest they were up and readying the spacecraft for a second plane-change to set up Yankee Clipper's orbit for a direct pass over the Fra Mauro landing site. Lasting 19sec, that SPS burn came at 159hr 04min 46sec for a ΔV of 382ft/sec (116m/sec) with bootstrap photography being the order of the day as the crew conducted a series of pictures which

would provide three-dimensional views enabling cartographers on Earth to produce contour maps along the projected descent paths.

For the fourth time, an Apollo spacecraft fired out of lunar orbit to return home, an event girded by anxious anticipation; with the Lunar Module gone, the burn back to Earth could only be performed by the SPS engine – it simply had to work to get the crew and their valuable collection of lunar material back home. And again, two clocks were running in the Mission Operations Control Room – one showing the time when the signal should be received as the spacecraft came in view again if the burn had happened as scheduled and another for 11min 18sec later indicating that Yankee Clipper would still be in orbit. Performed as always on the far side of the Moon, so as to raise apolune to beyond the equigravisphere and allowing it to be captured by the Earth's gravitational field, the TEI burn started when Yankee Clipper fired the SPS engine for 2min 10sec at 172hr 27min 17sec for a ΔV of 3,042ft/sec (927m/sec).

The coast home was largely uneventful, with six navigation sightings with some stars not in the Apollo star catalogue for determining the ability of a crewmember to get a fix on stars close to the Earth's bright limb. Unscheduled, but eagerly photographed by the crew was a solar eclipse which amazed and excited all three

RIGHT For Apollo 12 and 14 the Descent Stage oxygen quantity use was now required to support two periods of EVA, each time the LM was depressurised being shown by a sudden fall in the amount remaining. *(NASA)*

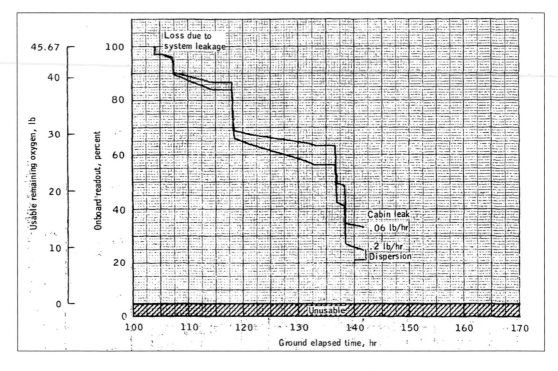

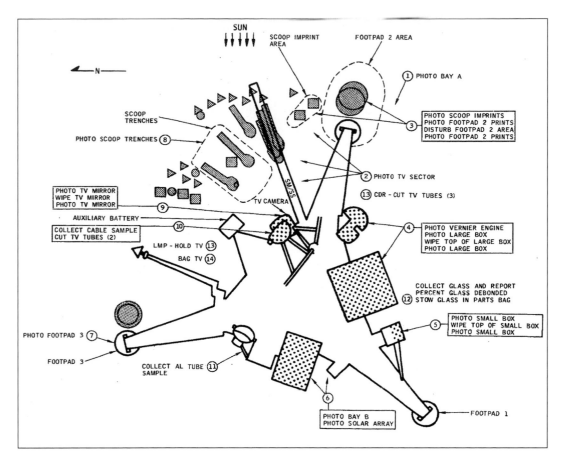

LEFT The plan to examine the Surveyor III spacecraft on EVA-2 required specified tasks to be completed as well as removal and return to Earth of selected items as shown here on this astronaut training chart. *(NASA)*

BELOW The second EVA took Conrad and Bean away from the Lunar Module and on a circuitous traverse around several station stops to the Surveyor Crater where the spacecraft sat awaiting inspection. *(NASA)*

crewmembers. The TEI burn gave an entry gamma (angle) of -7.24° to the local horizontal. On the way home two course corrections were made with the RCS thrusters, at 188hr 27min 16sec and 241hr 21min 59sec to adjust that to -6.48°, very close to the optimum -6.50° sought. Entry interface was achieved at 244hr 22min 19sec with splashdown 14min 05sec later into the Pacific Ocean.

But there was no soft landing for what remained of Apollo 12, the sea state being high and the impact 'extremely hard', which for a Navy crew was probably an understatement! Usually, the Command Module hangs suspended at an angle of 27.5° but in this instance it was at an angle of 20–22° which threw a pulse of 15g due to the wind swinging the spacecraft on its parachutes. The force even dislodged the 16mm data acquisition camera attached to a clip on the right-hand window, which came loose and struck Bean above the right eye. When the capsule was examined portions of the heat shield were seen to have been dislodged. Due to the heavy seas the Command Module flipped over to apex-

down on impact but was righted to the Stable II position by the inflatable 'golf balls'. Relaxed rules on biological isolation allowed the crew to wear blue coveralls and an oxygen mask over

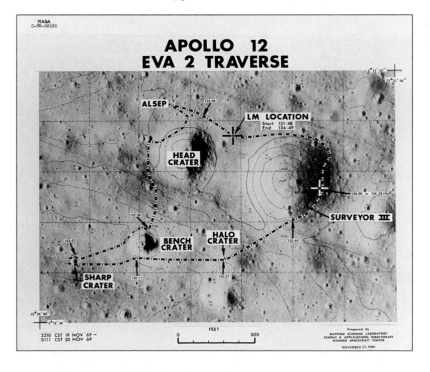

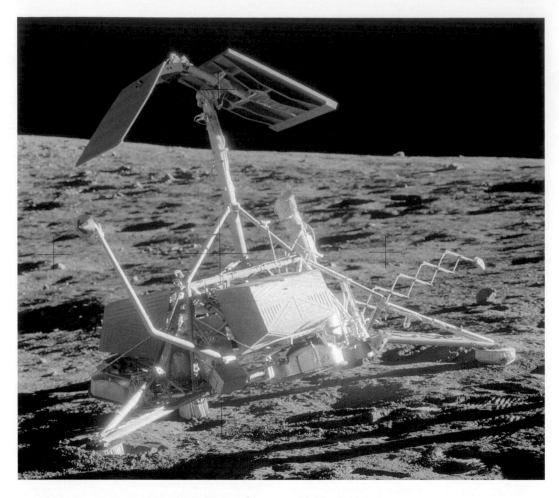

the facial area, although they were still taken to the Lunar Receiving Laboratory in the MQF van.

But reflections regarding the site they had visited added to the growing awareness of how places on the Moon differed from each other. Indicated by the amount of dust kicked up by the descent engine prior to landing, the site carried more fine particulate material than at the Apollo 11 site. On returning to weightlessness after lift-off, the crew were hampered by the quantity of this material floating around the cabin, making breathing difficult without helmets and affecting vision.

Surprising to the crew, the weightlessness of orbital flight lifted free dust from clogged cameras and the 'floating' effect removed large quantities of surface material which, on the Moon's 1/6th gravity, had thoroughly coated them. But the sheer quantity of dust was surprising and during the two lunar walks this factor caused serious problems with fastening glove and helmet locks. In fact, the ultrafine grains that got in to ring connectors lifted them free of a total seal, causing

oxygen leakage from inside the suits to come perilously close to redlines.

The rocks collected by the Apollo 11 astronauts had been a mixture of basalts and breccias, while the Apollo 12 rocks were almost all basalts, with only two breccias in the returned samples. Basalts are dark-coloured rocks which have become solidified out of molten lava. They are a very common type of volcanic rock on Earth and can be found in places such as Hawai'i. The basalts at the Apollo 12 site, however, are understood to have formed 3.1 to 3.3 billion years ago, which is approximately 500 million years later than basalts from the Apollo 11 site. Basalt consists primarily of the minerals pyroxene and plagioclase. At the Apollo 12 landing site, several different varieties of basalt were identified from the presence of other minerals such as olivine or ilmenite.

Overall, there was much less of the element titanium in the Apollo 12 samples than in the Apollo 11 samples, which explains the more

LEFT With the LM in the background on the crest of the crater's north west rim, Surveyor parts are retrieved and returned to Earth for analysis. *(NASA)*

BELOW Several decades later, NASA's Lunar Reconnaissance Orbiter took this view of the Apollo 12 landing site with foot tracks clearly visible as well as the Descent Stage of the LM and the ALSEP experiments. *(NASA)*

reddish colour of this region. The differences in age and chemical composition between the Apollo 11 and Apollo 12 samples demonstrate that Mare volcanism did not occur as a single, Moon-wide melting event. The Apollo 12 basalts formed from material that melted at depths of at least 93–155ml (150–250km) below the surface and then were brought up in lava prior to solidifying from where they were retrieved.

Examination of the Apollo 12 samples did confirm that some of the soil samples had come from Copernicus and this provided an age date of about 800 million years for that event. It came along with one unusual piece of material, a homologation of potassium, some rare Earth elements and phosphorus, carrying the acronym KREEP. Such rocks are enriched in potassium (denoted as K by chemists), rare earth elements (REE), and phosphorus (P). KREEP is believed to have formed early in the history of the Moon during the solidification of the Moon's molten stage, known as the magma ocean.

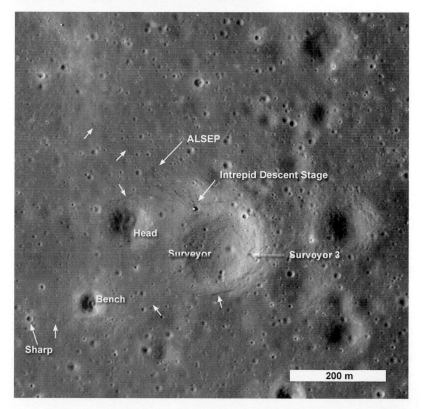

ALSEP

Intrepid Descent Stage

Head

Surveyor — Surveyor 3

Bench

Sharp

200 m

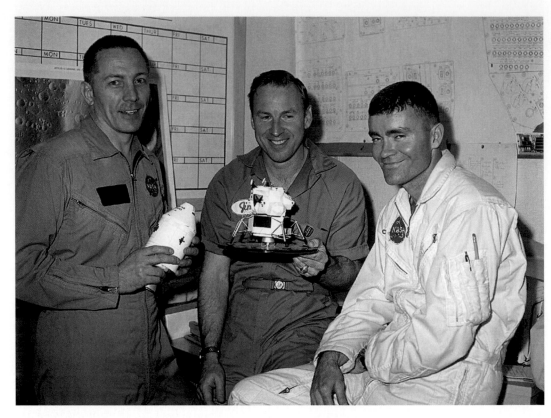

RIGHT Apollo 13 crew (from left) Jack Swigert, Jim Lovell and Fred Haise were launched on 11 April 1970 and experienced a total loss of power and oxygen from the Command and Service Modules, requiring them to use the Lunar Module as a lifeboat to get back to Earth after circumnavigating the Moon, the only Apollo mission to do so. *(NASA)*

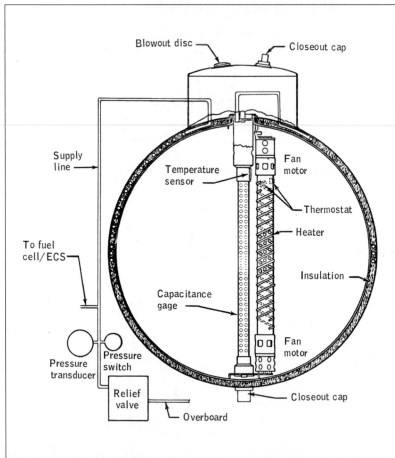

Truncated plans

With Apollo 12 successfully having demonstrated a pin-point landing in the Oceanus Procellarum the way was clear for flights to complex and difficult sites replete with geological treasures, albeit challenging to get in to. While retaining a notorious flexibility as to which specific site was assigned to a particular flight, on 29 July 1969, NASA had planned to fly the remaining missions at approximately five-month intervals. Following the Apollo 12 flight in November 1969, the remaining H-series flights (Apollos 13, 14 and 15) would fly in March, July and November 1970 respectively. These would be followed by the J-series missions (Apollos 16, 17, 18 and 19) in April and September 1971 and February, July and December 1972.

These plans were compromised by two separate issues, one regarding future NASA

LEFT One of the two cryogenic oxygen tanks exploded causing the second to leak to depletion but because there would be no EVAs from the Lunar Module, its oxygen supply was sufficient to keep three men alive on the return to Earth. *(NASA)*

programmes and another by the still declining NASA budget, down in this current financial year at $3.7billion from its peak of $5.2billion in fiscal year 1965. NASA hoped to reopen Saturn V production lines but negotiations with the Nixon administration for the fiscal year 1971 budget (beginning 1 July 1970) cancelled that option. After going through Congressional committees NASA received only $3.3billion for that year. Employment on all NASA programmes was already down from 450,000 to 190,000 and this caused a further 50,000 job cuts.

Because Saturn V was the only rocket capable of sending people to the Moon, the 15 already built was the limiting factor in continued lunar exploration and that ran headlong into another decision based on what NASA was planning after Apollo. Study contracts had already been awarded for designs on a permanently occupied space station, initially carrying 12 people but with the capacity to grow to 50 or even 100 personnel in a space base concept built up from several station modules. NASA wanted this space station programme operational in Earth orbit around the mid-1970s. NASA planned through its Apollo Applications Program (AAP) to launch a space station built around a converted S-IVB stage. Now, this AAP project was believed to be a suitable interface between the era of the Moon landings and the age of Earth-orbiting space stations.

However, when it became apparent in 1968 that this rocket stage converted into a habitable space could not operate effectively as a workshop in orbit, NASA began to look at a fully equipped S-IVB shell kitted out as a fully functioning space station and launched as such from the ground up, 'dry' versus a 'wet' concept. Instead of propelling itself into orbit as the second stage of a Saturn IB it would now have to be lifted by a Saturn V. When it became apparent during the autumn of 1969 that the proposed budget for fiscal year 1971 would be insufficient to order more Saturn Vs, NASA decided to cancel Apollo 20 and use the two stages of its Saturn V launch vehicle for sending what would soon be renamed Skylab into orbit, stretching out the Moon landings to six-month intervals.

This irked manned space flight bosses, who wanted to maintain momentum in launch processing and flight preparations and even Congress got in on the act, asking senior managers at the Marshall Space Flight Center just how low could the annual flight rate get before skill levels were lost and mistakes and errors began to creep in. The answer was two a year. NASA formally announced the inevitable end to Saturn V production, and the cancellation of Apollo 20, on 4 January 1970, six weeks after the Apollo 12 crew returned home and the day before the first Lunar Science Conference opened at the Lunar Science Institute in Houston.

Of course Apollo 20 was just a number but it represented the last of the five J-series missions. Now there would be four, following three more H-series flights, and because it would not be

BELOW The cryogenic oxygen tanks were located on the middle shelf in sector 4 of the Service Module, the outermost one being the one destroyed, shrapnel damaging the second tank seen here behind. Above is the base of the three fuel cells and below is the top of one hydrogen tank. *(NASA)*

exhaled carbon dioxide, each spacecraft used a uniquely-shaped lithium hydroxide canister which over time became absorbed and had to be changed. The limited supply of those in the LM ran out but the square-shaped canisters in the Command Module would not fit the drum-shaped orifice in the environmental control system. *(NASA)*

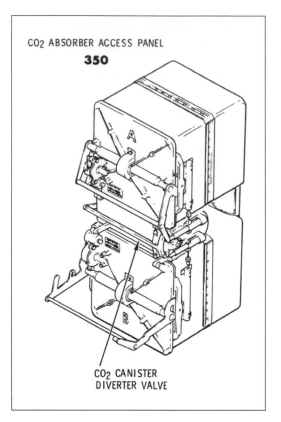

CO2 ABSORBER ACCESS PANEL

350

A

B

CO2 CANISTER
DIVERTER VALVE

BELOW Senior management are briefed by Deke Slayton on a workaround procedure whereby the square-shaped CM canister would be connected to the LM's environment unit. *(NASA)*

required NASA also cancelled LM-14 and the space suits and PLSS units supporting that mission. There was a sense of inevitability about that; NASA needed Skylab to bridge the gap between Apollo Moon missions and the new era of orbiting stations and a reusable launch vehicle known as the Space Shuttle, anticipated as a supply truck to the station and projected for operational debut along with the station. It had been at a meeting of the British Interplanetary Society in London on 28 August 1968, that NASA's manned flight boss George Mueller had first announced his plan for an era of reusability to cut costs and to sustain a fundable space programme for human participation.

But worse was to come, a political threat that many at NASA felt compromised their long-range plans. Although NASA had acknowledged that the Moon programme was limited in extent, and laid plans for the exciting and more advanced J-series missions (which see), only so much could be compressed into the remaining hardware and the agency was now running up against an uncommitted administration that saw the space programme as an indulgent drain on fiscal resources, questioning the wisdom of even the remaining Apollo flights and pondering the cancellation of all missions after Apollo 14.

Inaugurated as President on 20 January 1969, Nixon had never been keen on the human space flight programme and since occupying the White House had quietly questioned the need for NASA at all, asking advisers whether it could be adapted into a different agency for more vote-catching work. He feared the implications of a disaster upon his political reputation and was haunted by the prospect of losing lives on his watch.

Launched on 11 April 1970, Apollo 13 was unable to land at the Fra Mauro site when its crew, Jim Lovell, Jack Swigert and Fred Haise had to return to Earth following a near-catastrophe on the way to the Moon. One of two liquid oxygen tanks blew and disintegrated under excessive internal pressure, cracking the second tank which bled its contents to space. There was little option but to use the Lunar Module as a lifeboat and hopes of a landing were dashed. This undoubtedly fed into Nixon's lack of enthusiasm for the remainder of the Apollo programme and despite being unable to halt the progression of remaining flights, opposition both inside NASA and within Congress dealt Moon expeditions a final blow.

Opinions were divided at NASA between continuing with the lunar missions and cancelling one further expedition to use a Saturn V for a second Skylab station – a back-up was already planned. Or even yet another Saturn V for an advanced space station launched as a single structure. But why not cancel yet another, leaving only Apollo 14 and 15? This was precisely what was proposed in a report from NASA's Advanced Concepts and Missions Division at the Office of Advanced Research and Technology dated 6 April 1971. In this proposal, the last two Moon landings would take place in 1971 and 1972, leaving Skylab operations in 1972/73 and four 'interim' Skylab-type space stations between 1976 and 1983 so that the first two years of Shuttle operations would support the last two stations ready for the permanent station from 1987.

The contracted studies had always assumed that the permanent station would be sent to orbit by a Saturn V but that idea had died when it became clear that production of the giant rocket was over. And there were those who feared that the risks inherent with Apollo Moon missions outweighed the scientific value of lunar expeditions; what would be the repercussions, they said, of trying to convince Congress to fund a Shuttle and perhaps even a space station in the wake of a crew stranded on the Moon?

Despite these internal arguments, as always it would be down to what Congress was prepared to fund and, in the wake of a disinterested public rejection of TV shows from the Moon (the loss of the colour camera on the first EVA with Apollo 12 had not helped) and the general concern about safety, plus the feeling in political circles that Apollo had fulfilled its purpose in seeming to win the Space Race, there was little appetite for the level of funding required to sustain lunar expeditions. With NASA heading into a critical time for development of the Shuttle, it was time for risk-aversion and fiscal constraint.

With Congressional hearings for the fiscal year 1971 budget showing little stomach for giving NASA the money it wanted, by early spring 1970 there was already an inevitability

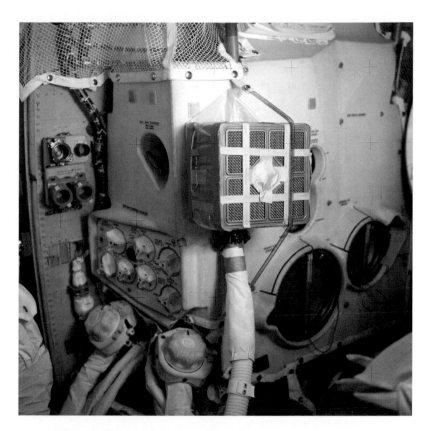

ABOVE Known as the 'mailbox', the jury-rigged unit adapted the CM canisters and when set up immediately began lowering the dangerous level of CO_2 which would have poisoned the crew. *(NASA)*

LEFT When the Service Module was separated shortly before re-entry, the crew took a picture of the damage exposed by the complete outer panel to sector 4 having been blown off during the explosion. *(NASA)*

RIGHT Several changes in senior management occurred after the Moon goal had been reached. In 1970, George Mueller left NASA as head of manned space flight, having successfully steered the agency on a track to the Moon after he was appointed in 1963. *(NASA)*

FAR RIGHT Former boss of the Apollo Command and Service Module programme at North American Aviation, in 1970 Dale Myers replaced George Mueller as head of the Office of Manned Space Flight. He would remain in position as NASA managed the remaining Apollo flights, the Skylab space station, the Apollo-Soyuz Test Project and the emerging Shuttle programme. Myers returned to Rockwell International in 1974. *(NASA)*

about the future and it was not in the direction of Moon missions. Still working on options, the space agency proposed a further compromise to adapt to what appeared to be a protracted migration to Shuttle/space station operations. Under a stretched-out schedule, two H-series flights would be flown in 1970 followed by one H-series and one J-series mission in 1971 and a second J-series in early 1972. Skylab would be launched later that year and visited by three sets of crews in 1973. This would be followed by the last two J-series missions in 1974.

As NASA worked through the modifications to the next spacecraft to fly to mitigate the risk of another 'Apollo 13', the decision was made and on 2 September 1970 the agency announced that two more lunar expeditions had been cancelled. Effectively, these were the two mission slots assigned to 1974. The remaining expeditions would be designated Apollo 14 through 17 but in fact the net result of these three cancellations was to remove one H-series mission (originally Apollo 15) and two J-series flights (originally Apollos 19 and 20) but to simplify matters they would now run consecutively by number. This left Apollo 14 as the last H-series flight before moving to the first J-series, the new Apollo 15. The year had been a gruelling one for NASA; buoyed with so

much success after the first two landings and now Apollo 13, triggering a year required to rebuild confidence back in the Apollo system while working the dramatic cutbacks which left scientists scurrying to find where to put the last four landings.

But that lay ahead. After a review board examined the reasons why Apollo 13 had nearly failed there were searching questions. It could have been avoided, said the inquisitors, if subcontractors had carried out work required by NASA instructions but never fulfilled. The oxygen tanks, already a flawed product and known to have unidentified problems, had been fitted with an electrical harness incapable of withstanding the higher voltages of ground support equipment at the Kennedy Space Center, which partially melted potting that caused an electrical short, heating the liquid oxygen into a gas beyond the capacity of the tank to withstand it and it failed.

NASA's engineering analysis provided a list of safety modifications which would provide better margins in the event of failure and improve the overall reliability of the system. The oxygen tanks were critical items – providing as they did not only the gas for crewmembers to breathe, but when mixed with hydrogen the electrical supply required to operate the spacecraft and water for the cooling system to thermally condition the equipment.

The oxygen tanks were redesigned with fan motors removed – these were originally designed to stir up the cryogenic oxygen to prevent stratification but operational experience showed that the fans were not necessary. The electrical leads were also to be encased in stainless steel sheaths with hermetically sealed heaters shielded from making contact with Teflon parts. A third heater element was also added to each tank so that each could be turned on sequentially, one, two or three at a time. In addition, a sensor was added to read the heater assembly temperature and the bulk temperature sensor was relocated within the tank to improve its accuracy.

A third oxygen tank was added to a bay in Sector I of the Service Module which was located on the opposite side of the spacecraft to the two existing oxygen tanks on the middle shelf in Sector IV. An isolation valve allowed the crew to completely isolate the new tank (No 3) from both the fuel cells and the other two tanks should that be required, feeding directly in to the environmental control system for breathing purposes. All three tanks had a new quantity gauge probe fabricated from stainless steel rather than aluminium and joints which had previously been soldered were replaced with brazed joints.

To improve the electrical power available, a 400-amp-hr silver oxide/zinc non-rechargeable battery identical to the four installed in the Lunar Module Descent Stage was located in Sector IV of the Service Module for supplementary or sole power in the event that the fuel cell system was lost. With the additional oxygen and electrical power the Apollo spacecraft was independent of installed conventional systems and could return the crew to Earth without recourse to the LM.

An emergency water supply comprised five, one-gallon (3.78l) plastic bags enclosed in beta-cloth and packed in a stowage bag with fill hose, valves and drinking nozzle. This stowage bag was attached to the aft bulkhead in the Command Module. In the event of a power-down to conserve electrical energy, that part of the spacecraft could conceivably freeze but prior to that occurring the bags could be filled and any remaining water just before re-entry could be dumped overboard through the waste management system.

There had been other changes, to the management structure at NASA, over the preceding 18 months. Jim Webb left NASA in October 1968 before the flight of Apollo 8 and on 10 December 1970 George Mueller left NASA and moved to General Dynamics, his place as head of manned space flight taken by Dale Myers from Rockwell (renamed from North American Aviation). Sam Phillips had departed NASA during the Apollo 11 flight in July 1969. On 1 March 1970, von Braun left his post as Director of the Marshall Space Flight Center and moved up to headquarters in Washington DC, where he was appointed Deputy Associate Administrator for Planning; he would retire from NASA on 26 May 1972, disillusioned with the retraction from deep-space exploration.

H-3 Apollo 14

Launch date: 31 January 1971
Duration: 216hr 01min 58sec

Selected for what was the last H-series mission, the importance of the Fra Mauro site was sufficiently high on the scientists' agenda that it was decided to send Apollo 14 to the place where Apollo 13 had been targeted. In general, the flight plan closely followed the same operational procedures, with a surface stay time of 33.5 hours and two EVAs, but this time of 4.25hr each based on the conservative use of consumables on the previous two missions. Originally scheduled for 1 October 1970, the mission was delayed while modifications were made to CSM-110; no modifications were required for LM-8.

The prime crew for Apollo 14 included Alan B. Shepard, Stuart A. Roosa and Edgar D. Mitchell. Shepard had flown a 15min ballistic flight to become America's first man in space when he flew the MR-3 flight on 5 May 1961, arguably the single enabling event that cleared President Kennedy to feel confident about proclaiming the Moon goal; a failure of America's first manned space flight may have scotched that decision for good. But Shepard had been grounded when he went down with Ménière's disease in 1963, affecting his inner ear and removing him from command of the first manned Gemini flight.

Shepard was appointed chief of the astronaut office in November 1963. Arguably the most ambitious member of the astronaut corps, Shepard got a reputation for his icy 'cool' and became notorious for his acerbic attitude but in his new desk job he threw himself into a wide range of activities including personal business investments and deals while simultaneously engaging with the emerging technologies of the Gemini and Apollo era. On the advice of astronaut Tom Stafford he went to see an otologist, privately and under a different name, who managed to conduct a surgical operation that cleared the medical problem and put him back in contention, but that brought raised eyebrows and verbal opposition from a new cohort of astronauts eagerly competing for seats on Moon missions, especially as there were only a limited number of flights remaining.

In the days of medical uncertainty regarding the overall effects of space flight on older people, several astronauts questioned whether Shepard was young enough to fly again. Born in late 1923, he was only 45 when Neil Armstrong (38) walked on the Moon. But dubious claims of his eligibility based on age were largely due to a determination to keep him off the roster for fear he would displace one of his accusers – which he did! Galvanised into a new fitness regime, the driving dynamic of Shepard's uncompromising determination pushed him relentlessly to lobby for a seat on a Moon mission and, despite some opposition at headquarters, he was eventually given command of Apollo 14 but only after being deferred from his attempt to get command of Apollo 13 on grounds that he needed the additional training that a shift to the next flight would allow.

It was a highly questionable assignment. Not only had Shepard not even made an orbital flight, he did not have the almost rigidly applied requirement to have served as a Command Module Pilot on an Apollo mission before upgrade to Commander. Not only that, despite his almost total lack of space flight experience, his other two crewmates were also making their first flight and this alone drew criticism among not just a few. For this reason, fellow astronauts watched as Shepard confounded their fears and excelled in the simulators and performed

BELOW The Apollo 14 crew was commanded by America's first astronaut into space, Alan B. Shepard, and included Command Module Pilot Stu Roosa and Lunar Module Pilot Ed Mitchell. *(NASA)*

well in training, pitching headlong and with enthusiasm to a new role as lunar explorer.

Most surprised of all at Shepard's assignment were the media representatives who for several years had been covering NASA human space flight activity and fully understood the rules and rationales – not all of them written down – which determined who did what and which crew got which mission. Nevertheless, there were just a few among the elite group waiting for a flight, especially the other six Mercury astronauts, who raised a metaphorical glass to one of their own who would make it all the way to the Moon and back.

As back-up, NASA chose Eugene A. Cernan (Gemini IX-A, Apollo 10), Ronald E. Evans and Joe H. Engle. With only Cernan having made previous flights, this crew would itself become the focus for attention when they came up for the last Apollo Moon mission of all, Apollo 17 (which see).

The landing site selected for Apollo 14 was in what selenologists formally called the Fra Mauro Formation and, specifically, close to what was informally named Cone Crater, with the primary objective of sampling material excavated by the Imbrium impact. Despite the successful pinpoint landing of Apollo 12, landing sites were still restricted to regions near the equator. Also, selected landing sites had to accomplish high-priority science objectives within the confines of two four-hour-long walking EVAs.

Fra Mauro material had already been mapped through Earth-based telescopes as being widely distributed across the nearside of the Moon. It therefore served as a convenient stratigraphic marker, dividing features that are older than the Imbrium impact from those that are younger. By returning samples of the Fra Mauro Formation for study on Earth, a precise age could be assigned to this geologic transition. Also, because the Fra Mauro was ejected by the Imbrium impactor, it was hoped

RIGHT Changes here identify improvements and safety features to significantly reduce the risk of another explosion by removing the very means by which that, and other risks might be encountered. *(NASA))*

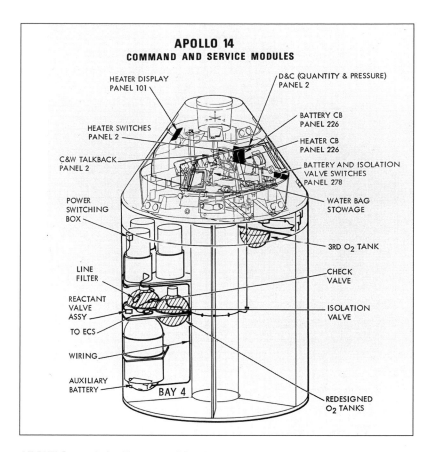

ABOVE Several significant modifications were made to the Block II spacecraft following lessons learned from the near failure of Apollo 13 to get its crew back home. An additional oxygen tank was located in the previously vacant sector 1 of the Service Module increasing total capacity by 50%, all oxygen tanks were redesigned and made safer and several technical and switching improvements were applied. These upgrades and improvements would remain a feature of this and the three J-series missions. *(NASA)*

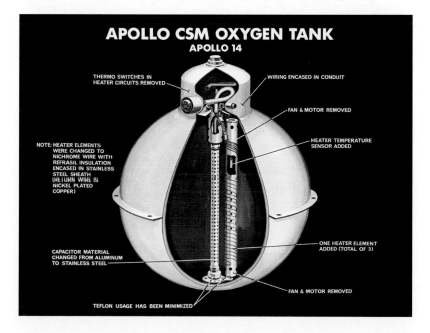

that it would provide samples that originated deep in the Moon's crust, perhaps from tens of miles (kilometres) below the surface.

The specific landing site within the Fra Mauro Formation was chosen to be near Cone Crater, a young, fresh, 1,214ft (370m) diameter impact structure, chosen because it is large enough to penetrate through the lunar regolith that has been deposited since the Fra Mauro Formation was formed. In a sense, Cone Crater served as a natural excavator, exposing the rocks that were the main objective of the landing. Prior to the abort of Apollo 13, Apollo 14 had been targeted to land in the Littrow region of Mare Serenitatis, where the objective was to study young, pyroclastic volcanic deposits. The Fra Mauro site was regarded as scientifically more important than the Littrow site.

Also, landing in Fra Mauro would allow the astronauts to obtain orbital photography of the Descartes region, something that was not possible if Littrow had been the landing site. Descartes was regarded as a high-priority target for a later mission (eventually flown by Apollo 16, which see), but could not be certified

as a safe landing site based on Lunar Orbiter photography. Although Littrow was rejected as the Apollo 14 landing site, another site in Mare Serenitatis, Taurus-Littrow, was later explored by Apollo 17 (which see).

Training for Apollo 14 had taken 20 months and the crew were advantaged by the preparatory work carried out by Apollo 13 which was expecting to go to the same Fra Mauro site. But the science equipment for Apollo 14 was different to that for Apollo 13, which had carried a drill for a deep-core sampling experiment while Apollo 14 carried an Active Seismic Experiment and a Modularised Equipment Transporter about which more later in this section. The crew selected the name Kitty Hawk for the Apollo spacecraft after the place in North Carolina where the Wright Brothers first took to the air, and Antares for the Lunar Module.

The launch was scheduled for 15:23:00hr local time on 31 January 1971 and the countdown and crew installation went on without significant issues but weather forced an unscheduled hold of 40 minutes beginning at T-8min. The launch window for a Sun elevation angle of 10.3° at the landing site had a total duration of 3hr 49min. The next window would not open until 15:03hr local time on 1 March. However, suspect clouds evaporated, lift-off of AS-509 occurring at 16:03:02hr local time necessitating a slight shift in launch azimuth from 72.067° to 75.558° so as to align the antipode point for TLI with the new position of the Earth–Moon geometry.

Following a standard Saturn V ascent, with early cut-off of the centre engines for both first and second stages, the spacecraft reached orbit and conducted two revolutions in essential checks and clearances for TLI, which occurred at 2hr 34min 32sec. The trajectory was updated so as to shorten the transit time to Lunar Orbit Insertion by 40 minutes, making up time lost in waiting for suitable weather and putting the relative position of the spacecraft back on the scheduled flight plan.

Separation of CSM-110 from the SLA on top of the third stage came at 03hr 02min 29sec but docking did not occur until 1hr 54min 27sec later with LM-8 extraction 50min 18sec after that. The extended delay was caused by a recurring problem with securing a hard dock

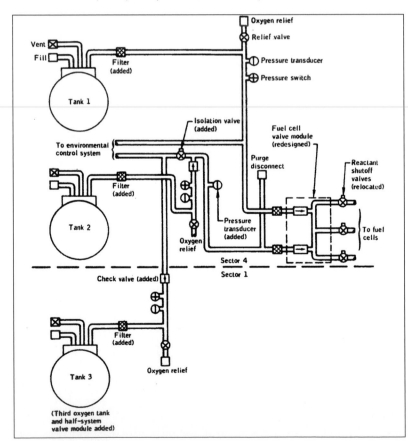

BELOW A schematic of the folding-in of the third oxygen tank to the existing plumbing and electrical wiring while at the same time allowing an option for isolation to prevent failures in any two tanks cascading to the third. *(NASA)*

with the Lunar Module. Roosa drove Kitty Hawk forward to engage but the three small capture-latches in the nose of the probe failed to snag with the receptacle on Antares. It was 3hr 14min into the flight and the first problem had arisen. Believing that he was doing too good a job at gently nudging up to the drogue on the LM, Roosa went at it a second time with more speed but again the latches failed to engage.

The third attempt had Roosa hold the thrusters firing the spacecraft forward for four seconds, pushing hard against the drogue. Still no luck, and again none on a fourth attempt when several scratches were noticed on the conical side of the drogue as viewed through the spacecraft windows. A fifth and final attempt was made and when contact between the probe and drogue was observed the probe was manually retracted with the spacecraft driving forward as though the latches had snagged. As the probe retracted the two spacecraft were driven together and achieved a hard dock when all 12 latches engaged and ripple-fired to their locked positions. Later, the tunnel was opened and the probe removed but when manually

activated inside Kitty Hawk the small capture latches worked as designed.

The first rest period began at 16hr 10min and ten hours later, shortly before reaching the halfway point in distance, Mission Control got a call from the spacecraft to say that they were awake. A debriefing ensued over the performance of the Saturn launch vehicle, standard procedure at this point in the mission.

ABOVE Preparing for flight, Shepard and Mitchell go through flight procedures in the simulator at the Kennedy Space Center. *(NASA)*

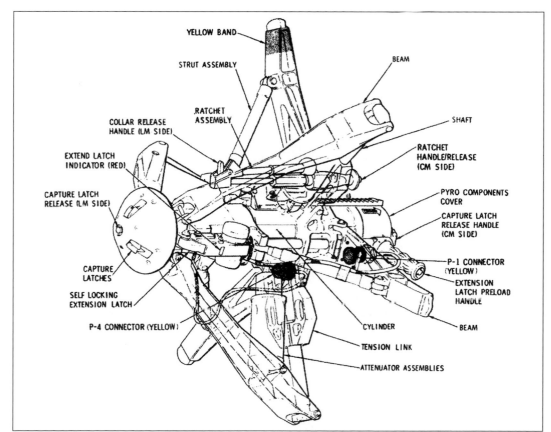

YELLOW BAND
STRUT ASSEMBLY
BEAM
COLLAR RELEASE HANDLE (LM SIDE)
RATCHET ASSEMBLY
SHAFT
EXTEND LATCH INDICATOR (RED)
RATCHET HANDLE/RELEASE (CM SIDE)
CAPTURE LATCH RELEASE (LM SIDE)
PYRO COMPONENTS COVER
CAPTURE LATCH RELEASE HANDLE (CM SIDE)
P-1 CONNECTOR (YELLOW)
EXTENSION LATCH PRELOAD HANDLE
CAPTURE LATCHES
SELF LOCKING EXTENSION LATCH
CYLINDER
BEAM
P-4 CONNECTOR (YELLOW)
TENSION LINK
ATTENUATOR ASSEMBLIES

LEFT After launch, problems were encountered with the docking procedure. This diagram shows the detail on the probe, which was designed to extend and project capture latches into the central hole in the docking cone on the Lunar Module. *(NASA)*

ABOVE The docking system was open beneath the Boost Protective Cover (BPC), which covered the conical Command Module to protect it should the Launch Escape System (LES) fire to blast the crew to safety in the event of a launch vehicle malfunction. The BPC came off when the LES was jettisoned, but water may have leaked through to the capture latches on Apollo 14 and frozen them open, thus preventing their actuation. *(NASA)*

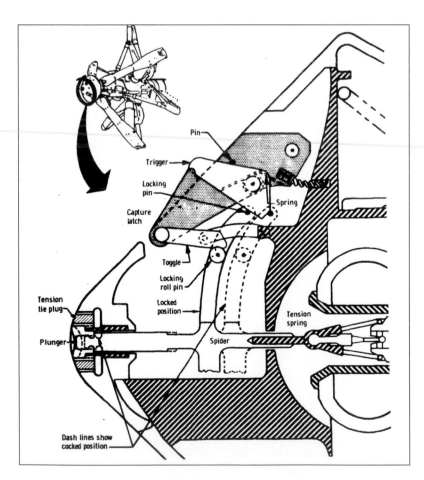

The TLI burn had put Apollo 14 on course for a free return pass around the Moon which would bring it no closer than 2,421ml (3,896km) but a hybrid transfer burn of the SPS engine at 30hr 36min 08sec reduced that to 77ml (124km). Another refinement of the trajectory would probably be needed. Post-separation manoeuvres using the Auxiliary Propulsion System on the S-IVB placed that stage on an impact trajectory with the Moon as planned, the stage striking the surface at an elapsed time of 82hr 37min 52sec and a velocity of 8,343ft/sec (2,543m/sec), at 8.07°S latitude by 33.25°W longitude albeit some 217ml (350km) off target.

At 54hr 53min 30sec the clocks in Mission Control were put back by 40min 03sec to synchronise the events with the actual arrival, which had been shortened to converge the time as well as the geometric position when the spacecraft arrived at the Moon. This was known as Phased Elapsed Time (PET). The clock aboard the spacecraft started at lift-off so it too had to be updated. It also avoided numerous updates to the Flight Plan on board and in Mission Control so that the GET in the checklists were synchronised.

Several hours later the crew took out the probe and drogue, opened Antares' hatch and inspected the interior of the Lunar Module while sending a 42-minute telecast to Earth. Antares was devoid of Mylar fragments or pieces of debris – although a washer had once floated out from behind a panel. But the astronauts, who were inspecting their LM for the first time after launch, did encounter five or six very small screws and washers drifting around weightlessly! Power came on to the LM at 61hr 41min 11sec and the only anomaly noted on the displays in Mission Control was a slightly low voltage reading on battery No 5. But there was nothing visible on either the docking probe contact switches or the drogue to suggest why they had such a problem docking after TLI.

After returning to Kitty Hawk, Shepard and Mitchell joined Roosa and another sleep session began. After waking up to the day they would enter lunar orbit, Mission Control advised of

LEFT A detail of the head of the docking probe with the capture latches identified. *(NASA)*

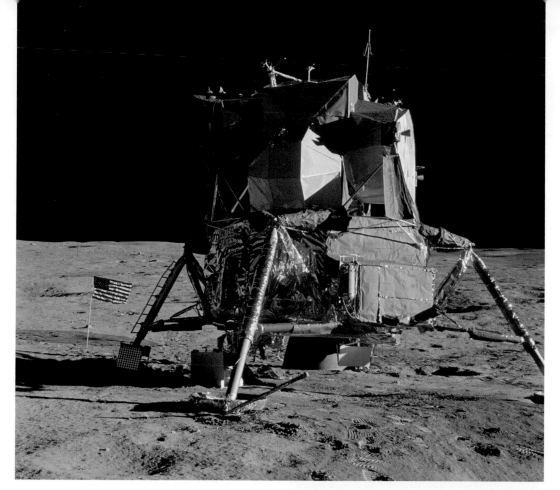

LEFT Lunar Module Antares on station at the Fra Mauro landing site. In clear view is the Scientific Equipment Bay containing the ALSEP array and the RTG cask on the left side of the bay. *(NASA)*

BELOW Mitchell takes a picture of Shepard out on the surface, shielding his face from the Sun, his identifying red stripe signifying that he is the Commander. *(NASA)*

a small tweak burn to the trajectory, a 0.6sec firing of the SPS engine to impart a velocity change of 3.3ft/sec (1m/sec) at 76hr 58min 11sec, lowering perilune to 67.5ml (112km). But the low voltage on battery No 5 worried engineers on the ground and Mission Control had Mitchell go into Antares and conduct a complex set of switching operations to obtain a health check; astronaut Joe Engle had worked the problem with specialists and voiced the procedure up to Kitty Hawk before handing back to Capcom Fred Haise.

But there was still concern over the docking probe anomaly and further word was passed to the crew on the close examination of spacecraft telemetry completed during the preceding hours. Nobody saw anything untoward about that and the final 'go' for continuing with the planned mission brought relief.

Under mission rules a wide range of seemingly minor problems could switch the mission to a contingency plan, of which there were many for each flight, and although it was quite possible to conduct an EVA to get from one spacecraft to the other, as demonstrated on Apollo 9, the assumption that the landing

ABOVE A view of the aft leg and the footpad which has dug in to the regolith. (NASA)

BELOW The Buddy Secondary Life Support System introduced to Apollo 14 was a precaution in the event that one backpack failed. By hooking up an umbilical the consumables from one astronaut could be supplied to the other but the requirement to ensure that sufficient remained to walk back to the LM restricted the distance from the Lunar Module. (NASA)

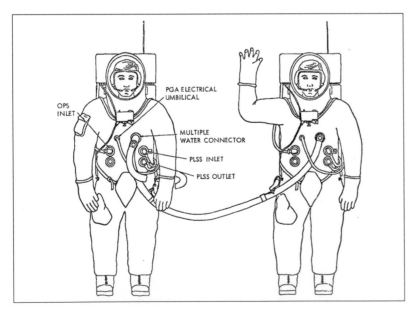

could go ahead with a failed docking system was by no means a certainty. But on this occasion there was nothing to say that a normal docking and undocking could not take place. One theory postulated that rain before the flight could have deposited water in the latch voids, freezing them in the cold of space; they were certainly working when manually depressed inside Kitty Hawk.

Around the Moon swept Antares and Kitty Hawk to fire the SPS engine again, this time in a long 6min 11sec burn beginning at 81hr 56min 40sec to put the docked vehicles into lunar orbit of 194.5ml x 66.8ml (313km x 107.5km) for LOI-1. It was a manoeuvre that had been conducted four times before but this time it was a little different. Instead of burning first into an elliptical path around the Moon, the SPS usually firing at the end of the second revolution for LOI-2 to place the spacecraft in a circular path, Apollo 14 would use LOI-2 to place itself in an elliptical orbit of 67.7ml (109km) by 10.5ml (17km) until the two vehicles separated on revolution 12, swooping low across the Moon's surface uprange of the landing site for more than 17 hours.

This manoeuvre combined the LOI-2 and DOI manoeuvres of previous lander missions and saved energy for the Lunar Module in that it was not required to place itself into the elliptical orbit prior to powered descent to the surface. That had several advantages. It allowed the CSM to conduct landmark sightings and to provide an extended period of tracking in the orbit the LM would use to begin PDI. Of course no orbit remains fixed and changes occur built on the heterogeneous structure of the Moon and the different gravity wells across the surface. For this reason persistent tracking allows a picture to build enhancing the accuracy of known parameters of the flight path – and that converges to provide still greater accuracy in landing. But it also saved precious DPS propellant which added to the available burn duration for precise touch-down selection with an added 15sec of hover time.

While the advantage of extended tracking time was a great benefit to the refinement of position and guidance equations, the combined LOI-2 and DOI manoeuvres conducted by a 20.8sec SPS burn at 86hr 10min 53sec had its risks. In lowering apocynthion of 194.5ml (313km) to a new pericynthion of just 10.5ml (17km) mean lunar radius there was the potential for an overburn which would bring pericynthion down below the surface. After the docked vehicles arrived around the eastern limb of the Moon, ground-based tracking verified that the manoeuvre had gone as planned; had it not there was just 12 minutes to perform a bale-out burn to lift pericynthion above the surface and prevent an impact.

Nevertheless, while there had been high confidence that the burn would have gone as planned, Mission Control began to fill up with managers and officials, astronauts and off-duty controllers for this was certainly a tense time. Not least in fact for the crew, giving the Command Module Pilot Stu Roosa a uniquely close view of the Moon which his predecessors had not experienced. But even the Lunar Module Pilot was impressed:

MITCHELL: 'Looks like we're getting mighty low here. It's a very different sight from the higher altitudes.'

CAPCOM: 'In about four minutes you'll be at your minimum altitude, which should be about 40,000ft (12,200m) above the terrain. We were wondering how things looked down there.'

MITCHELL: 'Well, I'm glad to hear we're that high. It looks like we're quite a bit lower – as a matter of fact we're below some of the peaks on the horizon, but that's only an illusion...The surface appears to be a lot smoother down here where we can see closer to the detail and particularly at this higher Sun angle it appears to be a softer surface, but it certainly is an unusual sensation flying this low.'

Photographic activity centred on the Descartes region, a future candidate landing site nestled in the craggy lunar highlands where old eroded craters and pockets of accumulated debris raised hazards and made pin-point landing a necessity for arriving at the right spot and essential for safety. But the integration of bootstrap photography and pre-mission reconnaissance heralded the upcoming J-series flights, adding much-needed high-resolution

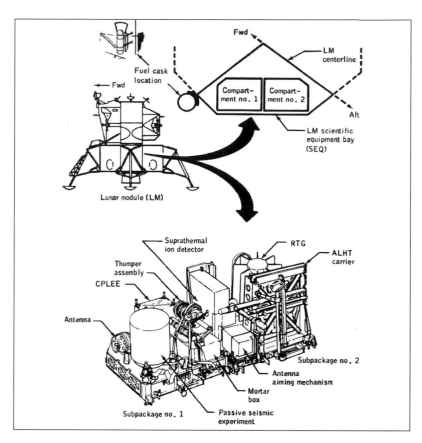

imagery of candidate locations for the three remaining expeditions. Orbital science would have to wait, however, as the crew needed sleep. The rest period prior to landing and EVA-1 events lasted about six hours with Mission Control back talking to the crew at 98hr 50min, all three having been awake for some time.

During the rest period taken by the crew prior to commencing the landing preparations, Apollo 14's orbit changed gradually as the asymmetric mass of the Moon pulled and tugged at the spacecraft's path. But all this had been factored in and the data from four previous lunar orbit missions played into the finesses essential for predicting how the mascons and minicons would alter the shape of orbits over time. As calculated, when the two spacecraft separated for the descent phase at 103hr 47min 42sec, pericynthion had decayed to 8.7ml (14km). But then a significant challenge to the landing emerged which delayed the descent.

At 104hr 16min 07sec a command to abort was detected at a computer input channel, despite the fact that the crew had not depressed the abort switch. If that abort command was present at the time the powered descent

ABOVE The ALSEP packages in the SEQ bay of the Lunar Module followed the same engineering design as that carried on Apollo 12, with the exception that the instruments were different. (NASA)

RIGHT ALSEP
subpallet No 1 with
affiliated equipment
including equipment
for the Active Seismic
Experiment. *(NASA)*

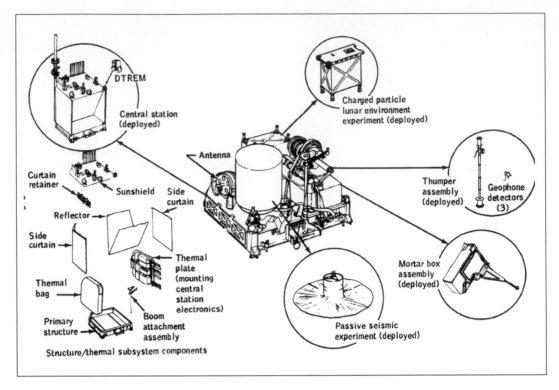

computer programmes were activated the PGNCS would immediately switch to the Abort Guidance System which would stage the LM and fire the APS to return the Ascent Stage to lunar orbit. The landing events could not proceed until the situation had been resolved and it took some time for a workaround to be prepared.

This consisted of four distinct parts, the first being entered into the computer immediately after the final attitude orientation for powered descent had been completed. It consisted of loading the AGS programme number into the mode register in the computer's erasable memory, which is used to display the programme number to the crew. While this did not instruct the active programme to change, it inhibited the computer from checking the abort command status bit. This also inhibited the automatic command to the full throttle position and automatic guidance steering and compromised processing of landing radar data.

To get around all these inhibiting consequences, the crew would manually control throttle-up at 26sec after engine ignition immediately followed by three further steps. First by setting a status bit to instruct the descent programme that throttle-up had occurred and that guidance steering was enabled. Second, resetting a status bit to disable the

abort programme. Third, replacing the active programme number back into the mode register so that the data from the landing radar could be processed by the computer after it had locked on to the surface. By doing all this the capability of the PGNCS to control an abort was lost and the crew would have had to manually switch to the AGS had an abort been necessary.

The crew had been alerted to this situation shortly after coming into view on the 13th revolution flashing up on Programme 52. The bit did not influence events only because the LM guidance computer was in a navigation alignment programme and a few quick checks revealed it to be a consequence of minute specks of dust trapped in one set of contacts in the abort switch.

To set up the geometry of the two spacecraft, Roosa performed a planned circularisation burn at 105hr 11min 46sec, settling Kitty Hawk into an almost circular path 68.4ml (110km) above the mean lunar surface and although this had taken the Apollo spacecraft into a path which would move it further behind Antares, it was in the correct orbit for rendezvous if the switch problem could not be solved. The time-urgency on this was that every orbit the Lunar Module made prior to landing was time eaten out of reserves in the consumables for supporting

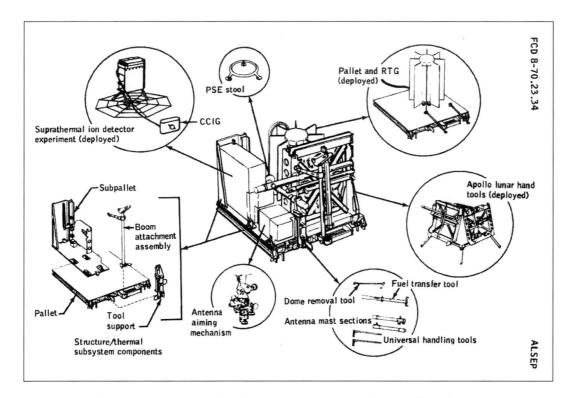

PSE stool

CCIG

Suprathermal ion detector
experiment (deployed)

Pallet and RTG
(deployed)

Subpallet

Boom
attachment
assembly

Apollo lunar hand
tools (deployed)

Pallet

Tool
support

Antenna
aiming
mechanism

Dome removal tool

Antenna mast sections

Fuel transfer tool

Universal handling tools

Structure/thermal
subsystem components

ALSEP

LEFT ALSEP subpallet No 2 supported the RTG, tools and some experiments as identified. *(NASA)*

operations with Antares and the orbit itself was steadily migrating west.

But it took time to find a solution and a lot of work with the software and computer specialists to find a way to safely prevent an inadvertent abort while protecting the option of a critical action should the need arise. Nevertheless, not until Antares came around the Moon on revolution 14 did the crew get the final procedures, a mere 35 minutes prior to PDI.

The descent programme was reselected in the primary computer just 10min before ignition, which occurred at 108hr 02min 27sec with the DPS at 10%, manually ramped up to 100% at 26sec. But there was also a problem with the steerable S-Band antenna and the crew were advised to use the omni-antennas to prevent a total dropout of voice and data. At ignition Antares was 304ml (490km) uprange of the landing site accompanied with a cry of 'Right on the money, right on the money'. The first computer entry to inhibit the abort command had been made just before the final trim and the remaining entries were made after the burn began. After the following throttle-up, the throttle was returned to the idle position, the burn now under PGNCS control.

About 42sec after the burn began the crew punched in computer guidance and a landing

point target of 2,800ft (853m) was entered at a burn time of 2min 15sec. Throttle recovery occurred approximately 12sec prior to the expected time but Mitchell read out the display data as Alan Shepard steered Antares down

BELOW The Central Station with sunshades shielding the equipment for connecting all the ALSEP instruments and the power sources as well as handling communications with Earth. *(NASA)*

RIGHT The Suprathermal Ion Detector is attached to the Cold Cathode Ion Gauge, both connected to the Central Station. *(NASA)*

BELOW The Laser Ranging Retro Reflector positioned so as to receive laser beams transmitted from Earth, returning them from corner-cube reflectors. *(NASA)*

to Fra Mauro but radar data failed to appear until Antares had descended to an altitude of 21,000ft (6,400m). 'Whew, that was close,' sighed Mitchell, for without comparative data the descent could not continue for long. Shepard was supposed to check manual control but the late radar data prohibited that and the AGS and the PGNCS closely matched at an altitude update at 12,000ft (3,650m).

Programme 66 went in at 8min 44sec and at an altitude of 8,000ft (2,440m) and the vehicle's forward pitch occurred as expected with the anticipated view instantly recognisable without reference to the Landing Point Designator. When referred to, they showed zero errors in crossrange or downrange. At 2,700ft (823m) Shepard redesignated the landing spot 350ft (107m) to the south to manoeuvre toward a smoother place.

But as Antares descended to around 1,500ft (457m) it became apparent that the redesignated spot was actually too rough and that the auto-land programme was taking the LM to a place short of the target. But Shepard was also aware that he needed to stay as close to Cone Crater as he could, for that was where their geological field trip would take them. Shepard opted for manual descent at 360ft (110m) and was brought to a hover at 170ft (52m) but there was no observed dust fanning out from below the LM obscuring the view. Forward motion was at about 27mph (43.4kph), reducing to zero over the landing point.

Then at 110ft (33.5m) the dust began to flow outward from the nadir position but Shepard found it relatively easy to control Antares and move with ease to his chosen spot. Using his acknowledged piloting skill, Shepard slowed the forward speed to a creeping motion until gently placing Antares on the lunar surface, noting that the fan of dust blown up by the descent engine was only about 6in (15cm) deep through which boulders and rocks were clearly visible. Seeking to achieve a gentle touchdown, Antares landed at 108hr 15min 09sec after a burn duration of 12min 44.6sec.

During descent the communications had been conducted on the forward omnidirectional antenna, switched to the aft omni at High Gate and then on to a fixed orientation for the

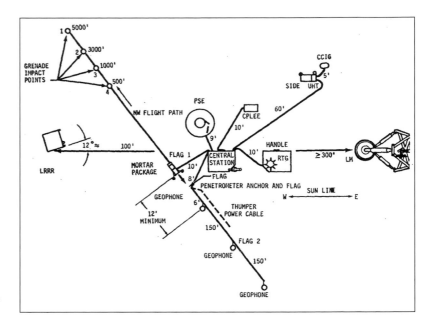

troublesome steerable antenna after Mission Control gave the 'go' for T-1 and then T-2. Using a gravity alignment with Programme P-57, conducting an optical alignment on specific stellar coordinates, Mission Control deduced that Antares was sitting on a slope of 6.8°, far below the 15° prohibiting a descent down the ladder or the 42° prohibiting a lift-off.

Apollo 12 had the advantage of two previous landings upon which to build a code of work which, while conforming to the Flight Plan and

ABOVE The ALSEP array showing the array of geophones for recording seismic waves caused by small charges detonated at the surface. *(NASA)*

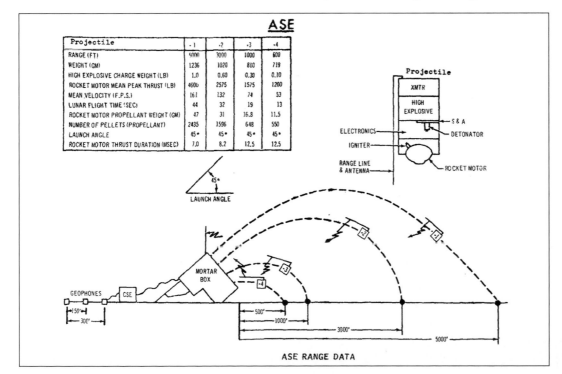

LEFT Larger charges lobbed from a mortar after the crew left the surface would help scientists profile the outer layers of the Moon. *(NASA)*

the checklists, also provided valuable orientation for the crew. Both Shepard and Mitchell found the post-landing description of the landing site, conducted in some detail, helpful not only for its obvious scientific contribution but also as a mechanism for giving the crew themselves a useful orientation concerning surroundings and familiarity with terrain which they found helpful for the upcoming Moonwalks. Preliminary analysis indicated that Antares was sitting only 150ft (50m) from its pre-planned spot and given the requirement for the Commander to select a specific area based on closer inspection while landing than any prior photograph could provide, was definitely a pin-point landing.

Preparations for EVA-1, again like Apollo 12 preceding the sleep period, went ahead well despite some troublesome issues with minor communication irritations regarding the PLSS antenna. But the first Moonwalk officially began at 113hr 39min 11sec, about 5hr 40min after landing. First order of business after getting down the ladder was to deploy a wheeled tool carrier folded up and attached to the exterior of the MESA in Quad IV of the Descent Stage. Called the Modular Equipment Transporter (MET), it consisted of a two-wheeled handcart upon which would be attached the Apollo Lunar Hand Tool Carrier used on Apollo 12.

The MET wheels were 16in (40.6cm) in diameter and 4in (10cm) wide inflated to 1.5lb/in² (10.3kPa) with a nitrogen gas and baked pre-flight so as to steam out antioxidants in the rubber. The MET had an empty weight of 26lb (11.8kg) and could carry up to 140lb (63.5kg), although on the lunar surface the effective loaded weight would be a comfortable 28lb (12.7kg). The low gravity of the Moon imposed a distinct 'bounce' effect when pulled along using a single handle in front. The MET would be used to carry two Hasselblad cameras, film magazines, weigh bags, sample bags and a novel feature for Apollo 14 – the Buddy Secondary Life Support System (BLSS).

Development of the BLSS was prompted by the extended range of traverse during EVA-2 on Apollo 14 compared with Apollo 12. Concerns about the ability of an astronaut to return to the LM in the event of a failure to his PLSS inspired development of the PLSS-6 pack, carried for the first time on this mission, in which a failed unit could share cooling water with the other astronaut's PLSS. It lightened the load on the Oxygen Purge System (OPS) which had been developed as a secondary system carried on top of the main PLSS for emergencies.

Under the 'buddy' system, the OPS would provide oxygen for breathing and suit pressurisation while the cooling water could be shared. It consisted of a single connecting pipe for water but was only carried on the MET because the astronauts remained relatively close to the LM on EVA-1. Nevertheless, the potential requirement to share cooling water influenced EVA planners in that the total remaining quantities factored in to the distance the astronauts were allowed to roam away from

BELOW Mitchell walks the line detonating charges from his 'thumper' device to set off seismic waves detected by the array of geophones. *(NASA)*

the LM so that they would always remain within walk-back distance sharing a single water tank.

Other changes included a different configuration of Sunshade on the helmet which allowed the wearer to adjust the areas shaded to the face without having to fully deploy a single visor conformal with the bowl-shaped faceplate. And the Commander would carry red stripes to identify the individual in photographs and TV images.

The ALSEP-C array for Apollo 14 included the Passive Seismic Experiment (PSE), the Suprathermal Ion Detector (SIDE) and the Cold Cathode Ion Gauge (CCIG), all of which had been carried on Apollo 12. Two new experiments were included for Apollo 14: the Charged Particle Lunar Environmental Experiment (CPLEE) and the Active Seismic Experiment (ASE). The crew would also set down a second Laser Ranging Retro Reflector (LRRR), the first of which had been deployed by Armstrong and Aldrin on Apollo 11, and conduct measurements with a Lunar Portable Magnetometer (LPM) on EVA-2. The CPLEE was designed to measure solar protons and electrons reaching the surface and would contribute information needed to better understand the interaction of the solar wind and auroras in the Earth's outer atmosphere. The insulated case weighed 5lb (2.2kg) and consisted of a small box 10.3in (26cm) by 4.5in (11.4cm) by 10in (25.4cm) containing two spectrometer cases, each of which had six particle detectors.

But the ASE was something very different. Seeking to obtain seismic profiling data with a graduated series of small explosive detonations, the ASE consisted of an array of three geophones placed at 10ft (3m), 160ft (48.8m) and 310ft (94.5m) from the Central Station. It would be used in conjunction with a 'thumper' device consisting of a pole with a cylindrical box at one end inside of which were 21 small explosive charges, each detonated on depression of a handle at the top. On walking back toward the Central Station the astronaut would detonate the charges at intervals of about 15ft (4.5m), the resulting shocks picked up by the geophones.

Scaling up the profiling for the PSE, a mortar device would be set up on the surface and activated about 10ft (3m) north-west of the Central Station and positioned so

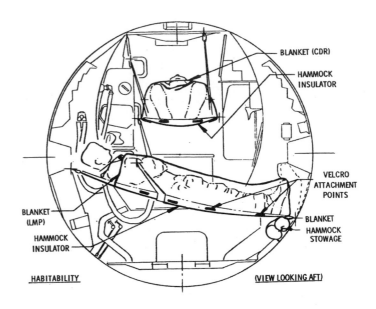

ABOVE The advised sleep arrangement in the Lunar Module involved an arrangement of hammocks as shown but these configurations were rarely used, each crewmember finding his own preferred way of resting. *(NASA)*

that when fired the mortar projectiles would land in a line directly opposed to the line of geophones, laid out in a south-east direction. Several months later these would be fired on commands from Earth, sending four grenades, at intervals from the mortar of 500ft (152m), 1,000ft (304m), 3,000ft (914m) and 5,000ft (1,524m). The thumper-geophone assembly

BELOW A bright Sun rising high for a second day exploring the Moon. *(NASA)*

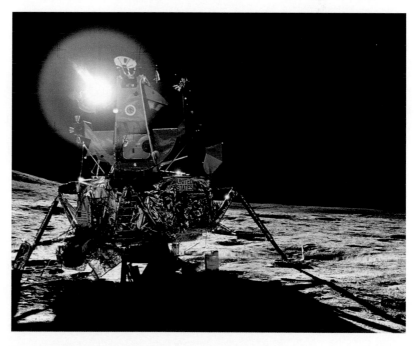

measured 44.5in (113cm) when deployed and weighed 7lb (3.17kg), with each geophone 4.8in (12.2cm) high, 1.6in (4cm) deep and weighing less than 1lb (0.45kg). The mortar package weighed 15lb (6.8kg) and consisted of a box 15.6in (39.6cm) in length by 4in (10cm) wide and 9.5in (24cm) high.

Employed only on EVA-2, the LPM was used for measuring variations in the lunar magnetic field at various places in what would constitute a relatively lengthy traverse toward Cone Crater and back. It was carried along on the MET and mounted on a tripod for use. The sensor head was attached to the data package by a 50ft (15m) cable and after placing it on the tripod at least 35ft (10.5m) away from the package the astronaut would return to the MET and provide a reading to Mission Control. Separation distance was important since the instrument was sensitive to the PLSS backpacks worn by the astronauts. Readings were made in three orthogonal axes and it was Mitchell's job to call out the measurements in each axis at one-minute intervals.

With the two astronauts out on the surface the first order of business after setting up the MET was to deploy the TV camera to a position about 50ft (15m) away from Antares.

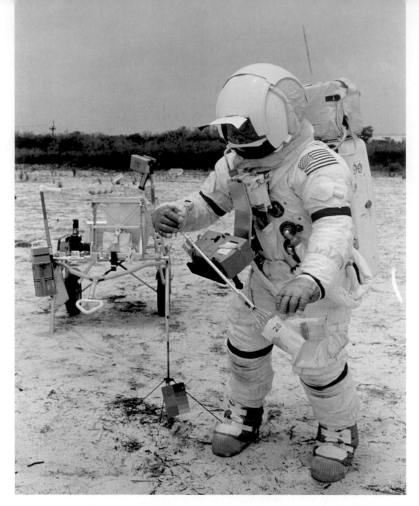

ABOVE The suit had significant improvements based on experience with two previous Moon landings and most of those were to the helmet Sun visor arrangement, displayed here with rehearsals for sample procedures. *(NASA)*

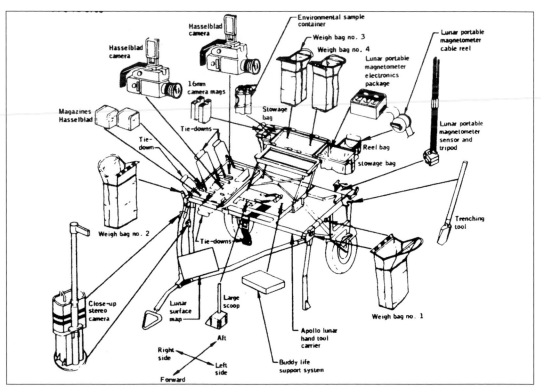

RIGHT The Modularised Equipment Transporter (MET) was the first wheeled vehicle on the lunar surface and was designed as a hand cart to carry tools, equipment, cameras, sample bags and a range of equipment each astronaut would need during the long geology excursion on EVA-2. *(NASA)*

Some 43min into the EVA the umbrella-shaped erectable S-Band antenna had been set up and live TV was possible after Mitchell switched the communication link, a picture showing the crew busying with duties around the LM. For this mission there was a back-up black-and-white TV camera – in case the colour camera failed and imposed another sound-only broadcast from the Moon.

Then Mitchell set up the LRRR and the Solar Wind Collector which would be retrieved and placed inside the LM at the end of EVA-2. Only a brief interval was devoted to raising the flag and temporary formalities conducted before the crew set about deploying the ALSEP instruments, described above. This was a lengthy process and the Moonwalkers fell behind schedule as they struggled with some of the equipment. About three hours into EVA-1, having been given a 30min extension on their 4hr 15min Moonwalk, Shepard was aligning the antenna on the Central Station and Mitchell set to work with a penetrometer. Pressing it into the surface he was able to get a measurement on the bearing strength of the regolith, the device easily reaching a depth of 19in (48cm) without encountering any noticeable subsurface obstruction at this particular site.

The ASE 'thumper' device failed to operate fully, only 13 of the 21 charges being detonated, five misfiring and three initiators deliberately not fired. But good readings were obtained at the PSE close by. Delays attempting to resolve the misfired charges slowed the crew a little further, ample demonstration that not all activities carefully worked out in 1g back on Earth could necessarily be trouble-free on the lunar surface. In no small measure, that was one of the functions of conducting work on the Moon, determining the effectiveness of tools, equipment and procedures. Mindful of the considerable expansion in surface activity on the J-series missions, engineers and technicians scrutinised every procedure to see what could be learned, how equipment could be improved and what was necessary to increase the chances of success where, in broad terms, surface work for each mission would increase from 16man/hrs during two EVAs on the H-series missions to 48man/hrs on the three EVAs of each J-series expedition.

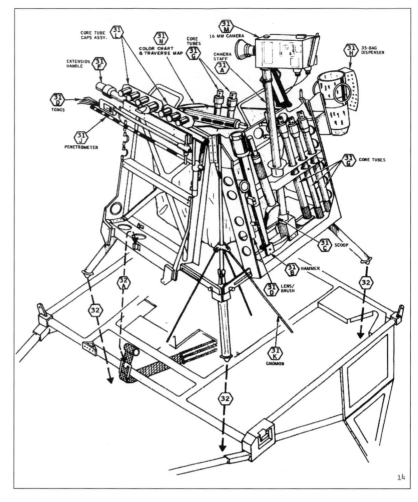

Not least among parameters being closely monitored were the consumables in each PLSS and at 3hr 37min Shepard had 50% oxygen remaining, Mitchell 34% but the timeline had drifted to the right on the activities-achieved chart and there was less time for sample collection when they returned to Antares. At the end of EVA-1, Shepard had consumed 53% of available oxygen, 56% of feedwater and 81% of available battery power; Mitchell's figures for each consumable were 78%, 66% and 84%, respectively. EVA-1 ended at 118hr 27min 01sec, a duration of 4hr 47min 50sec. After almost 22.5 hours, Roosa began a sleep period in Kitty Hawk at around 121hr 15min, his day like that of Shepard and Mitchell extended by delays to getting down on the surface. But it was a further 90 minutes before the two Moonwalkers began what was scheduled as a seven-hour rest for them, which would have had them awake at 128hr 45min.

Long before that, Antares put in a call to

ABOVE The Apollo Lunar Hand Tool Carrier could be placed on top of the MET as shown. *(NASA)*

ABOVE A '1g' MET
trainer was used
to familiarise the
astronauts with its use
and versatility. (NASA)

movement forward the surface becoming more undulating, the secondary impact craters increasing in frequency and the boulders getting bigger. Shepard collected samples at this site and, gaining ground on their objective, the crew rested to plan their assault on the rim of Cone.

At 2hr into the EVA the two astronauts were entering the southern boulder field of Cone Crater and the going was getting more difficult. Mission Control requested another rest stop and decided to halt attempts to go on. Shepard resisted, and on they went, climbing up slopes of 10–15°, negotiating boulders 10–12ft (3–3.6m) high, and all the while pulling the MET along with them. Shepard's heartbeat had climbed to 150/min, Mitchell's up to 128/min and again they stopped for a rest. Huge, half-buried slabs were to left and right; large angular blocks were randomly distributed all across the slopes, and jagged boulders of numerous shapes and sizes blocked the view forward.

At this point Shepard voiced the opinion that in his estimate it would take another 30 minutes to reach the rim of the crater and showed concern that they would absorb all their time on traversing the boulder field and have little time left for sampling. But, at 2hr 4min, Mission Control gave a 'go' for a 4hr 45min EVA and Mitchell persisted in his determination to press on, Shepard conceding.

It was probably some 10min later that the two astronauts came closest to the rim of Cone Crater. Later, photographs would show that they were just 75ft (23m) south-west of the rim proper. Working westward now, they began to collect samples, documenting as they went, driving core tubes into the surface and taking readings with the portable magnetometer. With heart rates down to an acceptable 108/min for Shepard and 86/min for Mitchell, Mission Control advised them to leave the area at 2hr 40min.

Their plan was to work down the south flank to a site called station D, and then on to Outpost Crater in the valley south of their outward trek, station E. Revising this, they were asked to proceed all the way back to a crater called Weird – named for its appearance – designated station F, about 1,000ft (300m) east of Antares. With 45% of oxygen remaining in Shepard's PLSS and 40% in Mitchell's, they

Mission Control at 126hr 20min, with a jolly 'OK, we're up and running this morning.' Scurrying along to build back time already eating into their stay on the Moon, Shepard and Mitchell were intent upon getting a day's work in on the surface and receiving a 30min extension to their planned 4hr 15min activity. EVA-2 formally began at 131hr 08min 13sec with Shepard going out the door followed by Mitchell, the intent of this expedition being to conduct an extensive field geology survey, with the MET carrying a lunar surface camera, four sample weigh bags, two 70mm cameras, two magazines, two 16mm camera magazines, gas analysis sampling equipment, a trenching tool together with the equipment supporting the LPM.

Field stations along the assigned route were denoted by letter (A to H, some with subsets). The first station, A, had the crew leaving the vicinity of Antares at 30min into the EVA and passing a small group of craters on the way called Triplet. They took geological photographs, drove core tubes into the surface and performed a thermal degradation experiment to aid in the selection of paint constituents for the Lunar Roving Vehicle (LRV) planned for the three J-series missions. At 1h 15min the two explorers headed due east to station B, more than halfway to Cone Crater and set on a gently rising slope, with each worked their way down the undulating and rocky slope, across the Mare blanket to Weird Crater. Arriving some 20min later, the two men remarked on the difficulty of moving down-Sun with their long shadows blacking out the forward path. While documenting samples at station F they were advised to press on to North Triplet and perform all their traverse activities within a reasonable distance of Antares.

Stopping one crater diameter from this site, station G, they took a triple core sample, calling it off when repeated hammering failed to penetrate deeper than 20in (51cm). Shepard attempted to trench deep into the surface and found the sides continually caving in. Leaving the trench to obtain samples, he soon returned to obtain a collection of sub-surface fragments. These were important, material that could have been covered for several hundred million years. Leaving site G at 3hr 40min, the two explorers made their way back to the LM, arriving there 15 minutes later.

Bringing a lighter tone to proceedings, Shepard produced a couple of golf balls and with the extension handle, proceeded to tee off in the general direction of Fra Mauro. Ever the golf addict, and with a nod to his local club, Shepard made the most distant golf gesture ever attempted. But it was not at all clear whether he actually hit the balls or not!

The EVA ended at 135hr 42min 54sec for a duration of 4hr 34min 41sec and while the field geology trip had, overall, been an outstanding success, there were lessons learned, not least about the intractability of working on the Moon, an airless world without the light distribution properties of an atmosphere. And there were interesting lessons too for mobility across the surface.

In traversing the hilly region to and from Cone Crater there were occasions when the MET was carried between both astronauts, the energy required by one man to pull it over rough ground being excessive and mitigated by two. Overall, the MET performed exactly as expected in its role as the first wheeled vehicle on any other celestial body. The crew observed that there was almost no dust or surface material thrown up by the wheels, neither was there any build-up on the fenders, guards placed over the circumference of the upper wheel section

to prevent sprays of dust becoming attached to equipment, instruments or either astronaut.

For the next several hours Shepard and Mitchell stowed loose equipment and prepared for the ascent. They powered up the Lunar Module and configured the guidance systems for the lunar orbit rendezvous, but here again there was a difference from preceding flights because the method of rendezvous had been significantly modified, shortening the time to reach the CSM

ABOVE Shepard pauses to take a double core tube sample. *(NASA)*

BELOW Mitchell consults his traverse map before starting off with Shepard for Cone Crater. *(NASA)*

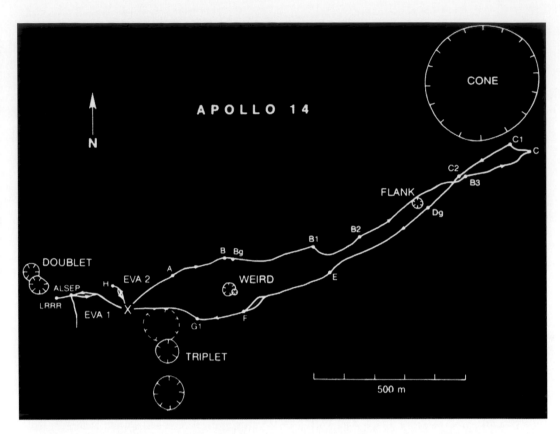

and dock, saving precious time for building in
a full-duration surface exploration and reducing
the length of the crew day, which had already
involved EVA-2. To set up for that direct-ascent
rendezvous, Kitty Hawk had performed a plane
change, an 18.5sec burn of the SPS inducing
a 370.5ft/sec (113m/sec) velocity increment, at
117hr 29min 33sec while Shepard and Mitchell
were nearing the end of their first EVA.

Rehearsed on Gemini rendezvous flights,
the direct-ascent technique relied upon precise
knowledge of the position and state vector of
each spacecraft as well as the tracking capacity
of the equipment which was feeding Mission
Control with parameters vital for precisely
targeting the Ascent Stage as it homed in
on the Apollo spacecraft. If successfully
demonstrated, this technique would be used
on the last three Moon missions and shorten
the time from lift-off to docking from 3.5hrs
to just over 1.75hrs. While still requiring a
margin should a more protracted rendezvous
be necessary as a result of anomalies, the
basic mission energy budget was reduced and
additional time gained for surface activity.

Lift-off occurred at 141hr 45min 40sec,
completing a 33.5hr stay at the surface. The
APS burn lasted 7min 12sec with low velocity
residuals trimmed out by the RCS thrusters.
The PGNCS, AGS and powered flight processor
showed very close comparators but a vernier
adjustment manoeuvre was performed 3min
57sec after the APS shut down. This was a
10.3ft/sec (3.14m/sec) RCS burn lasting 12sec
to place the position, rather than the velocity
(targeted by the ascent programme), in the

correct phase with the target up ahead. At that point the Ascent Stage settled into an orbit of 58.9ml (94.8km) by 9.6ml (15.4km).

Because of the increased speed of the Ascent Stage as it closed on the Apollo spacecraft the TPF manoeuvre was performed with the APS, the first time a second firing of this engine had been incurred on a Moon mission. This was a 3sec burn at 142hr 30min 51sec effecting a velocity change of 85ft/sec (26m/sec) around on the far side of the Moon putting the stager in an orbit of 69.2ml (111.3km) by 52.9ml (85.2km).

Although Kitty Hawk was unable to get good VHF ranging data until TPF had been accomplished (which occurred at 143hr 13min 29sec), the solutions converged closely with ground projections based on tracking from Earth. In fact, the ranging data showed a 10ml (16km) error which was not borne out by the measurements achieved through the sextant, because it prioritised VHF over visual tracking that did little to compensate for the error. Nevertheless, the two mid-course corrections satisfactorily resolved the apparent 'error' and a final rendezvous was achieved on time with docking at 143hr 32min 51sec, a mere 1hr 47min 11sec after lift-off and with no recurrence of the problems that had proved so troublesome in getting the two docked after TLI and transposition several days earlier.

Prior to docking, Kitty Hawk conducted a full 360° pitch manoeuvre for inspection of exterior surfaces, Shepard and Mitchell transferred to the Command Module and undocking occurred at 145hr 44min 58sec with a CSM separation burn less than a minute later to increase the distance between the two when the de-orbit manoeuvre was performed, an operation similar to that conducted by Apollo 12. The four thruster quads on the Ascent Stage fired at 147hr 14min 17sec in a 1min 16sec firing and the stage impacted the surface 28min 06sec later at 3.42°S latitude by 19.67°W longitude. This was 41.4ml (66.6km) from the seismometer at the Apollo 14 site and 71.3ml (114.8km) from the Apollo 12 site, a mere 8ml (12.8km) off the pre-planned spot.

De-orbit of the Ascent Stage occurred less than one hour before TLI, which began with ignition of the SPS engine at 148hr 36min 02sec for 2min 29sec, adding 3,460ft/sec (1,055m/sec) and propelling Kitty Hawk out of lunar orbit and on its way to the equigravisphere and capture by Earth. In an ebullient expression of precision, retro-officer Charles Dietrich calculated that this mission would set a record for accuracy, the burn having accomplished a predicted entry gamma of -6.81°, incredibly close to the optimum -6.5°, mid-point between a 2° corridor of acceptability for the flight path angle at entry interface. It would require a very small tweak, only a modest velocity change to counter dispersions in the trajectory incurred by waste water venting producing a propulsive offect on the flight path. That burn took place

BELOW Very near the rim of Cone Crater, the explorers pause and take a photograph, the MET appearing in view at bottom right. (NASA)

'Contact rock'　　　　　　'Saddle rock'

at 165hr 34min 57sec, lasting 3sec for a 0.5ft/sec (0.15m/sec) velocity adjustment nudging a -6.63° entry gamma.

But Command Module Pilots on every mission liked to show their skill at conducting a manual plot of their course, vindicating the expectations of Dr Charles Stark Draper several years earlier when designing the guidance and navigation equipment, creating the very concept of cislunar tracking. So it was that Stu Roosa began a concerted effort in the closing hours at proving his capabilities, manually obtaining a predicted entry gamma of -6.1° and then, after greater persistence, a value of -6.39°, within 0.05° of the final figure calculated by Mission Control, and an accuracy of 0.46ml (0.7km) of the vacuum perigee point.

With a range to go of 1,300ml (2,107km) some 75.75ml (121.8km) above the Earth, Apollo 14 encountered entry interface at 215hr 47min 45sec and a velocity of 36,170ft/sec (11,025m/sec), splashing down in the Pacific Ocean at 216hr 01hr 58sec. Visible to the crew on the recovery vessel USS *New Orleans*, the spacecraft was readily accessible to recovery divers and the crew were on deck 47 minutes after splashdown. NASA was excited to point out that despite problems with initial docking to extract Antares from the S-IVB stage after TLI, and a short in one of the circuits that delayed the landing, at 13 separate junctures in the mission an automated robot would have failed.

Four lunar orbit experiments had been conducted on Apollo 14, one of which, the S-Band transponder experiment, was designed to detect variations in the lunar gravity field caused by mascons and minicons but problems with the high-gain antenna made it almost impossible to get effective results. In a bistatic radar experiment the VHF and S-Band transmissions from the Apollo spacecraft provided data to assist with measuring surface roughness and depth of the regolith covering the outer surface of the Moon.

Somewhat more exotic was a search for the cause of a bright background light in the night sky opposite the side facing the Sun when

ABOVE Saddle Rock marks the closest the duo got to the rim of Cone, virtually right at the very crest of the depression. *(NASA)*

BELOW Weird Rock, where Shepard and Mitchell stopped for a station sample. *(NASA)*

viewed from the surface of Earth. Known as the gegenschein, this was believed to be a scattering of interplanetary dust focused at a point about 1 million miles (1.6 million km) from Earth and scientists asked the astronauts to make three separate sets of photographs from their displaced location so as to narrow down the options for its cause. The fourth experiment involved very close microscopic examination of the Command Module windows so as to determine the cratering flux of minute particles using images obtained before and after the flight.

Mission photography was accomplished from the Apollo 14 Command Module, which spent 67 hours in orbit both during DOI for high-resolution mapping photography and during its circular orbit for routine scientific photography; from the LM, which successfully landed in the hilly upland region 15ml (24km) south of the rim of Fra Mauro Crater; and by the Apollo 14 astronauts during extravehicular activities on the lunar surface.

The Apollo 14 crew returned 1,328 frames of 70mm photography and 15 exposed magazines of 16mm film. The lunar topographic camera (LTC) malfunctioned and only 193 usable photographs were recovered from the two rolls of 5in (12.7mm) film. A total of 17 stereopairs of lunar surface rocks and soil were taken on the mission and these enabled further study by those interested in lunar soil formation, impact phenomena, and soil mechanics.

The orbital photographic and scientific experiments conducted by the Command Module Pilot simultaneously with the surface exploration included the gegenschein photography and bootstrap photography using the Hycon and Hasselblad cameras. The bootstrap photography was accomplished using the 70mm Hasselblad camera with a 500mm lens after the failure of the Hycon lunar topographic camera. By using the crewman-optical-alignment-sight manoeuvre to hold the camera on target, some 40 stereo photographs of the Descartes landing area were obtained. Three passes were made over the Descartes area on revolutions 27, 28 and 30 to obtain stereo strips covering that region.

Most of the 92.6lb (42kg) of rocks and soil collected on Apollo 14 were breccias (rocks that are composed of fragments of other, older rocks). The countless impacts that have sculpted the Moon's surface broke many rocks down into small fragments. The heat and pressure of such impacts can sometimes fuse these fragments into new rocks, called breccias. In some cases, the rock fragments that form a breccia are themselves breccias. Such rocks obviously have experienced complex histories with multiple generations of impact events. Some breccias were heated to a point that some of the material in the rock was melted. Deciphering the histories of these rocks was, and still is, a great challenge.

Many of these breccias include material enriched in KREEP, first discovered on Apollo 12 (which see) but much more abundant in Apollo 14 samples. Although breccias dominate the Apollo 14 samples, some basalt samples were collected, usually as clasts (fragments) in breccias. Generally, the Apollo 14 basalts are similar to basalts found in Mare regions studied on other Apollo missions. However, the Apollo 14 basalts are generally richer in aluminium and sometimes richer in potassium than other lunar basalts. The Apollo 14 basalts formed 4 to 4.3 billion years ago, older than the volcanism observed at any of the Mare locations studied during the Apollo programme.

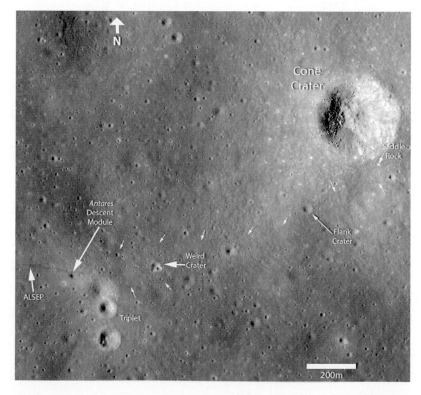

BELOW An image of the Apollo 14 site showing key locations where samples were obtained and providing visual testimony to how close they got to Cone Crater. The wheel tracks of the MET can be seen leading from the LM to Cone and back. *(NASA)*

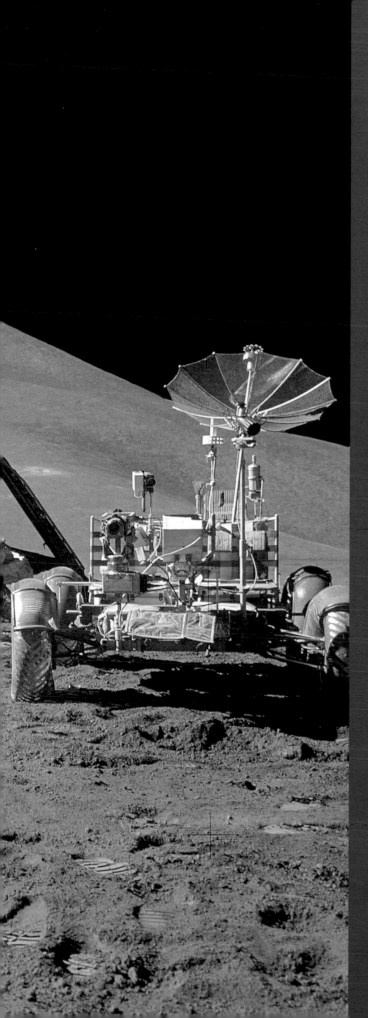

4 J-Series Missions

Completion of the two H-series landings cleared the way for the definitive development in Apollo hardware, setting up the scientific investigation of the lunar surface in a way that had been dreamed of by scientists and engineers for a decade since the dawn of the Apollo programme but impossible to achieve with the standard hardware as built for the political objective of landing on the Moon by the end of the 1960s.

OPPOSITE With Falcon on its perch, the MESA deployed and the LRV extracted and operating, Jim Irwin gives a salute on the plain at Hadley Apennine. *(NASA)*

113

It required a considerable expansion of capability and an advanced level of operational support. Where the H-series missions had 'grown' some Apollo capabilities following verification of the engineering potential by the G-series Apollo 11 mission, the last three Moon landings would go just a little way to showing what would have been possible with the lunar missions component of the Apollo Applications Program.

During the design of the Apollo spacecraft, and before it was picked up and applied to the Moon goal, provision had been made for a series of applications, including the carriage internally of scientific instruments for use in Earth orbit and around the Moon. These included the attachment of a space telescope in the Service Module that could be operated by the crew in the Command Module. As noted at the beginning of this book, the initial function of Apollo was to conduct deep-space operations and circumlunar flights and, sometime in the 1970s and supported by a landing system with legs, make the first lunar landing. Now that original design would support a greatly expanded role for the CSM in conducting lunar science.

Command and Service Module Upgrade

The basic design of the drum-shaped Service Module had six pie-shaped segments radiating out from a central tunnel which contained the Service Propulsion System and pressurising helium tanks. The

RIGHT The advanced J-series flights would each take advantage of an Extended Lunar Module, carrying a Lunar Roving Vehicle for transporting astronauts across the surface of the Moon. *(NASA)*

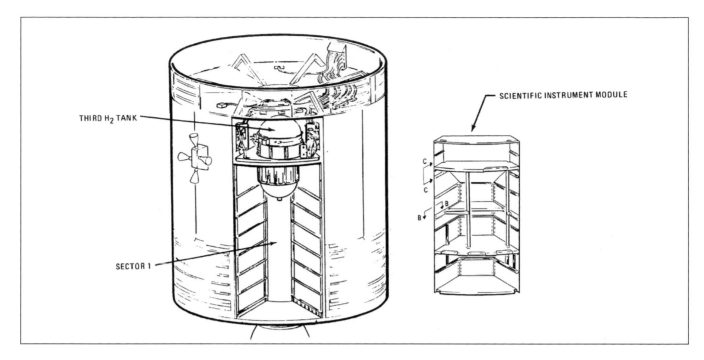

THIRD H₂ TANK

SECTOR 1

SCIENTIFIC INSTRUMENT MODULE

allocation of equipment was configured so as to leave vacant one segment (No 1) while dedicating another (No 4) to the fuel cells and associated oxygen and hydrogen tanks. The other four segments housed the propellant tanks for the SPS engine. Not all segments were equal – the propellant tanks were in 60° or 70° segments while the vacant sector and the fuel cell sector were 50° in radius. It was always the intention that the vacant segment would be the one to house scientific equipment including that telescope that never did get applied but which was carried forward as a concept to the Skylab programme – itself a product of the Apollo Applications Program.

The origin of the J-series really goes back to 1963 when NASA was expecting to have expanded lunar surface activity, a story for another time in another book, but began studies on what could be done with the basic hardware to extend the time on the Moon and develop a mobility system for roaming considerable distances from the Lunar Module on scientific expeditions. This was the basis of what became the AAP activity but the upward trajectory of expectations and planning collided with the downward path of the budget, ironically from 1965. Nevertheless, what the J-series represented was the best that NASA could get in the wake of collapsing funds.

ABOVE The advanced J-series flights would each take advantage of an Extended Lunar Module, carrying a Lunar Roving Vehicle for transporting astronauts across the surface of the Moon. *(NASA)*

BELOW The J-series Block II Service Module also contained a third hydrogen tank, increasing by 50% the cryogenic consumables installed before Apollo 14. *(NAA)*

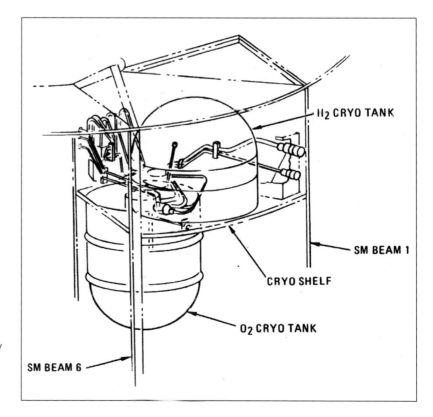

H₂ CRYO TANK

SM BEAM 1

CRYO SHELF

O₂ CRYO TANK

SM BEAM 6

The basic objective for each J-series CSM would include six days in lunar orbit out of a total mission duration of more than 12 days and conduct detailed scientific investigations of the Moon from a battery of instruments and equipment installed within the previously vacant sector of the Service Module in addition to the provision for carrying and deploying a small satellite released during the flight to remain in lunar orbit. In that vacant sector North American Rockwell (as NAA had been renamed) would install the Scientific Instrument Module (SIM), a box-like wedge-shaped structure containing all the additional equipment weighing more than 900lb (408kg) and with an internal volume of 300ft³ (8.49m³), some 25% more than the internal volume of the Command Module.

The SIM was 9.4ft (2.8m) tall and 5ft (1.5m) wide at the front, closed by aluminium honeycomb panels between 1in (2.5cm) and 1.5in (3.8cm) thick, its side and back walls consisting of aluminium panels strengthened by horizontal and vertical aluminium stiffeners. The SIM was shaped to fit the original sector and the back was blunted by the back wall consisting of the SM's central tunnel to a width of 2ft (0.6m). At a depth of 22in (55.9cm), the top one-third of the SIM was shallower than the bottom two-thirds, which was 44in (111.7cm). This provided room behind the shallow portion for a third oxygen tank, first installed on Apollo 14 (which see) after the problems with Apollo 13. To make room for the tank the back wall of the SIM was canted forward at an angle 35° from the horizontal between points 70in (177.8cm) and 93in (236cm) above the floor of the SIM. The wall then continued vertically to the top of the SIM.

The SIM was attached to radial beams 1 and 6 by 0.25in (0.6cm) bolts and to the new cryogenic tank panel and to the aft bulkhead of the Service Module. Three shelves in the SIM were similarly manufactured from a bonded aluminium sandwich structure, the lower one located 30in (76cm) above the floor, a middle shelf 26in (66cm) above the lower and a top shelf 37in (94cm) above the middle one and 20in (51cm) below the top. Protective covers and thermal blankets provided thermal control for individual instruments and the inside surfaces were coated with a material having an absorptivity/emissivity ratio of 0.05/0.4.

Two hollow, rectangular-shaped aluminium beams ran vertically from the lower to the upper shelf with horizontal fittings at middle-shelf height connecting the beams to the SIM walls. Inconel steel bolts were used to attach equipment, bolt holes being strengthened with aluminium spacers, flanged tube-like hollow rivets with the bolt head and nut acting as the closing force. Although appearing to be an integral part of sector No 1, the SIM was thermally isolated from the Service Module by minimising structural contact wherever possible and the only contact was at the attachment points along the front edges of the SIM wall and the radial beam caps on the SM. Gold tape was used on the narrow confines between the SIM sidewalls and the SM radial beams, both wall and beam faces being coated with the tape.

A jettisonable door protected the equipment until about 4.5hrs prior to Lunar Orbit Insertion. The door was a reworked version of an aluminium honeycomb Service Module sector panel, 9.4ft (2.8m) tall, 5ft (1.5m) wide and weighing 170lb (77.1kg). It was jettisoned by detonation of an explosive pyrotechnic train behind its edges and ordnance boosters at the four corners.

This train was composed of a single strand of an explosive cord similar to that used to separate the Apollo spacecraft from the SLA after TLI. The cord was positioned so that it would not leave any rough edges that could tear the space suit worn by the Command Module Pilot who would perform a deep-space EVA to retrieve film cassettes from the SIM; because the Service Module was destroyed during re-entry, those had to be re-located to the Command Module prior to re-entry.

Six ports in the SIM door and five on a smaller door above it provided pre-launch access to installations in sector No 1. The second door was a structural unit bolted around the edges and it was not jettisonable. It also consisted of a reworked section of an aluminium honeycomb Service Module door and was about 22in (55.9cm) high. It afforded access to the cryogenic oxygen and hydrogen tanks and to plumbing and cabling.

SCIENTIFIC INSTRUMENT MODULE ORBITAL SCIENCE PAYLOAD

The SIM payload for J-1 (Apollo 15) and J-2 (Apollo 16) missions included a mapping camera and film cassettes, a panoramic camera and film cassettes, laser altimeter, particles and fields subsatellite, gamma ray spectrometer, mass spectrometer and a combined alpha/x-ray spectrometer. Instead of these last three spectrometers, J-3 (Apollo 17) would carry an ultraviolet spectrometer, an infra-red scanning radiometer and a radar for a lunar sounder experiment in the space previously carrying a subsatellite. The actuators, gear boxes and wiring were applied to the different instruments and the controls remained in the same position in the Command Module.

Considered by many scientists to be one of the most outstanding advantages of the J-series flights, the 24in (61cm) Itek panoramic camera was 'gifted' to NASA by the US intelligence community. It had been developed from the IRIS II aerial reconnaissance camera which was an improved and redesigned version of the KA-80 which had been produced for the Lockheed SR-71 spyplane and had been carried on some of the U-2 spy flights over denied territory. Nothing much was made of the association, certainly not in official US government literature. But the pictures it would return on the film cassettes retrieved by the CMPs on the three J-series missions were far better than anything seen by the public before.

The camera was designed to produce stereo and high-resolution photographs with a published resolution of about 3ft (1m) but the originals would reveal detail down to a resolution of about 12in (30cm). The camera produced an image covering an area of 17.3ml (27.8km) down-track by 207ml (333km) cross-track with a field of view of 11° by 108° respectively. It obtained photographs by means of a rotating lens system which could be stowed for protection from thruster exhaust efflux. The film cassette weighed 782lb (32.6kg) and had capacity for 1,650 film strips. The panoramic camera was operated in conjunction with the mapping camera together with the laser altimeter. In this way it was possible to obtain a comprehensive map of 8% of the lunar surface, or 1.16million ml² (3million km²).

The 3in (7.6cm) Fairchild mapping camera was designed to provide photographs with a 65.6ft (20m) resolution on 3in (7.6cm) film together with a stellar camera shooting a star field on 35mm film at the same time at an angle of 96° from the optical axis of the surface camera. This system allowed correlation of the mapping photographs by using the photographs of the stars to determine the specific location on the surface of the vertical shots. The stellar shots also allowed position vectors for the laser altimeter during night side passes. The 3in (7.6cm) mapping camera had an f4.5 metric lens with a 24° field of view covering a square area on the surface of 106ml (170.5km) on a side from an altitude of 69ml (111km).

BELOW The Scientific Instrument Module in the configuration for Apollo 15 and Apollo 16. *(NASA)*

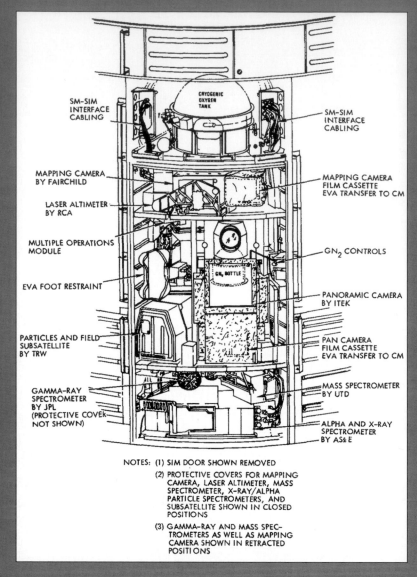

SM-SIM
INTERFACE
CABLING

CRYOGENIC
OXYGEN
TANK

SM-SIM
INTERFACE
CABLING

MAPPING CAMERA
BY FAIRCHILD

MAPPING CAMERA
FILM CASSETTE
EVA TRANSFER TO CM

LASER ALTIMETER
BY RCA

MULTIPLE OPERATIONS
MODULE

GN₂ CONTROLS

EVA FOOT RESTRAINT

PANORAMIC CAMERA
BY ITEK

PARTICLES AND FIELD
SUBSATELLITE
BY TRW

PAN CAMERA
FILM CASSETTE
EVA TRANSFER TO CM

GAMMA-RAY
SPECTROMETER
BY JPL
(PROTECTIVE COVER
NOT SHOWN)

MASS SPECTROMETER
BY UTD

ALPHA AND X-RAY
SPECTROMETER
BY AS&E

NOTES: (1) SIM DOOR SHOWN REMOVED

(2) PROTECTIVE COVERS FOR MAPPING
CAMERA, LASER ALTIMETER, MASS
SPECTROMETER, X-RAY/ALPHA
PARTICLE SPECTROMETERS, AND
SUBSATELLITE SHOWN IN CLOSED
POSITIONS

(3) GAMMA-RAY AND MASS SPEC-
TROMETERS AS WELL AS MAPPING
CAMERA SHOWN IN RETRACTED
POSITIONS

ABOVE This layout shows the arrangement of instruments and equipment for the first two J-series missions. *(NASA)*

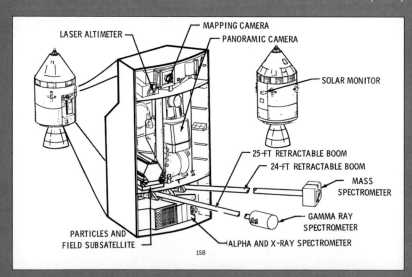

LASER ALTIMETER

MAPPING CAMERA

PANORAMIC CAMERA

SOLAR MONITOR

25-FT RETRACTABLE BOOM

24-FT RETRACTABLE BOOM

MASS
SPECTROMETER

GAMMA RAY
SPECTROMETER

PARTICLES AND
FIELD SUBSATELLITE

ALPHA AND X-RAY SPECTROMETER

158

The stellar camera incorporated a 3in (7.6cm) f2.8 lens with a 24° field of view with cone flats. The cassette for the mapping camera weighed 23lb (10.4kg) and had capacity for 3,600 frames which, with the stellar camera film, would be recovered by the CMP during trans-Earth coast.

The laser altimeter was designed to obtain a precise measurement of the distance between the orbiting CSM and the surface below to an accuracy of 3ft (1m). It was boresighted with the mapping camera and with the panoramic camera. When the mapping camera was operating the laser fired a pulse to align the mid-frame range to the surface for each photograph. The light itself was a pulsed ruby laser operating at 6,493angstroms and 200millijoule pulses of 10nanosec duration. It could repeat at up to 3.75 pulses per minute.

The subsatellite would be ejected into lunar orbit carrying three experiments: an S-Band transponder which would be used for gathering information about the lunar gravitational field and especially useful with the disruptive mascons and minicons; an experiment for determining the fields and particles encountered in the vicinity of the Moon as well as providing more detailed information on the Earth's magnetosphere; and a magnetometer experiment to measure the electrical properties of the lunar environment together with determination of the plasma interaction with the Moon.

The subsatellite itself was housed in a box-shaped container, spring-ejected out of orbital plane at a departure rate of 4ft/sec (1.2m/sec) and with a spin rate of 140rpm. The hexagonal subsatellite itself was 31in (78.7cm) long, with a width of 14in (35.5cm) and a weight of 78.5lb (35.6kg). Three booms, each 5ft (1.5m) in length and attached at 120° intervals around the structure, were deployed to their raised position after release. Electrical power was provided by an array of

LEFT This diagram shows the booms deployed for spectrometer sensor heads and the general configuration of the subsatellite which would be deployed on Apollos 15 and 16. *(NASA)*

solar cells across each flat surface providing 25W for operation on the day side and a rechargeable silver-cadmium battery for night side operations.

The gamma ray and mass spectrometers were mounted on extendible booms which were deployed 26.9ft (8.2m) and 25ft (7.6m) respectively from their stowed positions. The booms were formed by two curved, tempered steel tapes, wound flat on two motor-driven reels prior to deployment. As the tapes were unreeled they assumed their natural C-shapes and joined each other to form a tubular boom. Due to the strength and weight considerations of the booms they had to be retracted before SPS engine burns. If they were unable to be retracted the activating mechanisms, booms and spectrometers could be jettisoned by a spring-powered mechanism.

The gamma-ray spectrometer was carried along to measure the chemical composition of the lunar surface and was operated in conjunction with the X-ray and alpha-particle spectrometers to build an integrated compositional map of the lunar surface ground track. It was capable of detecting natural cosmic rays, induced gamma radioactivity with the lit and unlit sides of the Moon. An important part of the data gathering was to deploy this instrument on the way to the Moon and after the SIM bay door had been jettisoned. This was so that the instrument could measure cislunar radiation and gain a measure of the cosmological gamma-ray flux. The device was also operated so as to measure energy levels in the range of 0.1–10million electron volts.

The second instrument designed to obtain geochemical information about the Moon, the X-ray fluorescence spectrometer, was built to detect X-ray fluorescence induced by an interaction with solar x-rays and the Moon on the sunlit side. This would also be operated during cislunar coast as it shared the same compartment on the lower shelf covered by a protective door which could be opened from the Command Module cabin.

The alpha-particle spectrometer would measure alpha-particles from fissures in the lunar crust as a product of radon gas emitted by radiogenic compounds inside the Moon itself. The sensor operated in the detection

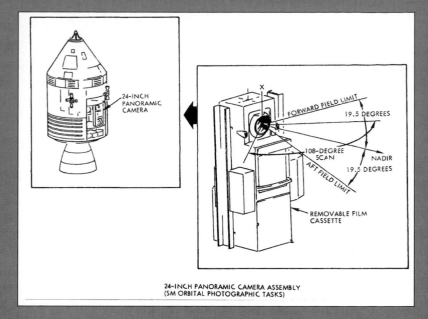

24-INCH PANORAMIC CAMERA ASSEMBLY
(SM ORBITAL PHOTOGRAPHIC TASKS)

ABOVE The operating envelope of the Panoramic Camera with angular image capture in each frame. *(NASA)*

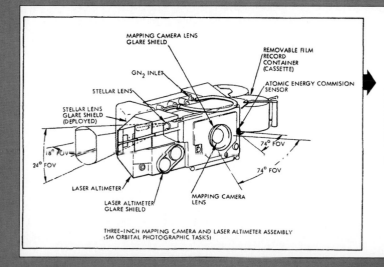

THREE-INCH MAPPING CAMERA AND LASER ALTIMETER ASSEMBLY
(SM ORBITAL PHOTOGRAPHIC TASKS)

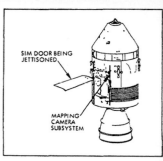

LEFT The Mapping Camera assembly, stellar camera and laser altimeter. The SIM bay was covered by a single panel and jettisoned prior to the spacecraft entering lunar orbit, as shown at right. *(NASA)*

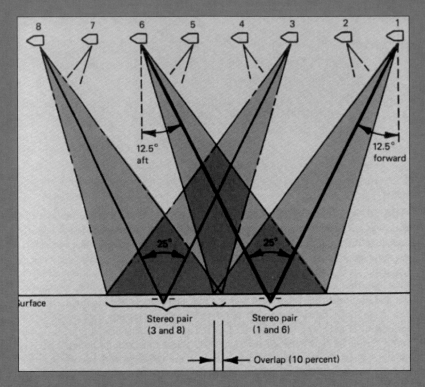

range of 4.7–9.3 million electron volts and was made up of 10 silicon surface barrier detectors. Like the other two spectrometers, this instrument was designed to operate along the orbital ground track but was not constrained by solar illumination. It was also to be used for establishing background alpha-particle emissions and it too was protected by a door, operated from within the Command Module.

The mass spectrometer was designed to measure the ambient lunar atmosphere rather than the surface conditions. Investigations would focus specifically on the lunar terminators at sunset and sunrise, the times when a range of gases were believed to be concentrated above the surface and through which the spacecraft would pass. The belief that the Moon has no atmosphere is highly relative and although concentrations were believed to be of very low density, trace gases are expelled from surface fissures, and the

ABOVE This diagram shows the geometry of exposures, with the roll frame rocking back and forward 12.5° in each direction. The solid lines show the centre of each frame converging at an angle of 25° for stereoscopic imaging. (ITEK)

RIGHT An example of the Panoramic Camera's forward and aft-looking frames which were set out for the Apollo 15 mission. (ITEK)

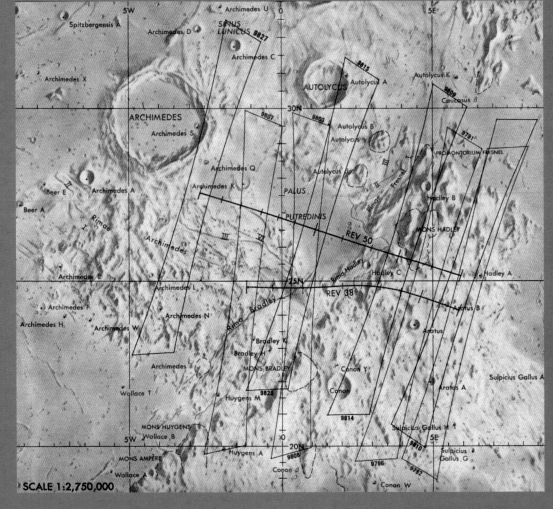

instrument was designed to measure species from 12–28 AMU (atomic mass units) for ions and 28–66 AMU with a second counter. The sensor head was mounted to a 24ft (7.3m) long boom. Under ideal situations, a continuous measurement over five lunar orbits was recommended.

These instruments were carried on flights J-1 and J-2 but J-3 (Apollo 17) incorporated a different suite of instruments as noted earlier. While retaining the panoramic camera, the mapping camera and the laser altimeter, the J-3 SIM bay replaced the three spectrometers and subsatellite with a far-ultraviolet spectrometer, an infrared scanning radiometer and a coherent synthetic aperture radar (CSAR), which incorporated a lunar sounder. The CSAR occupied the space where the subsatellite container was located for the J-1 and J-2 missions, while the cameras remained in the same positions as they were for the preceding three flights.

The far-UV spectrometer was carried along on the Apollo 17 spacecraft to detect and measure the atomic composition of trace constituents of a lunar atmosphere including far-UV radiation from the Sun reflected off the lunar surface. It was also expected to record UV radiation emitted by sources in the Milky Way galaxy. Built to gather data in the range of 1175–1675 angstrom, it was expected to record hydrogen, carbon, nitrogen, oxygen, krypton and xenon, although the quantities were expected to be in the billionths of those densities measured in the Earth's atmosphere.

The instrument itself was mounted on the lower shelf in the SIM bay and had an external baffle to limit stray light, a 20in (51cm) focal-length Ebert mirror spectrometer with 0.2in x 2.4in (0.5cm x 6.1cm) slits, a 4in x 4in (10cm x 10cm) reflection grating, a scan drive mechanism to provide the wavelength scan and a photomultiplier tube to measure the intensity of the incident UV radiation, together with the processing electronics. The spectrometer itself has a 12° x 12° field of view aligned 18° to the right of the SIM bay centreline and 23° forward of the vertical axis of the CSM, with all the operating controls installed in the Command Module for activation by the astronauts.

To obtain a lunar surface temperature map of higher fidelity than anything achieved thus far, the Infrared Scanning Radiometer (ISR) would obtain temperature measurements of the far side as well as the near side with a resolution of better than 1.8°F (1°C) and with a surface pointing location accuracy of 1.2ml (2km). This level of accuracy would enable scientists to accurately locate hotspots and to correlate observations of gases or visual obscuration that might indicate outgassing from the subsurface. Measurements obtained in this way would help establish thermal curves relating the surface temperature to the position of the Sun. These measurements would even help define the structure of the lunar crust and when combined with photographic data could assist with identifying local rock fields below the resolution of the images.

The ISR was mounted to the bottom shelf of the SIM bay, the instrument suite

BELOW This diagram shows the orientation of the mass spectrometer, deployed from the SIM bay on a boom. *(NASA)*

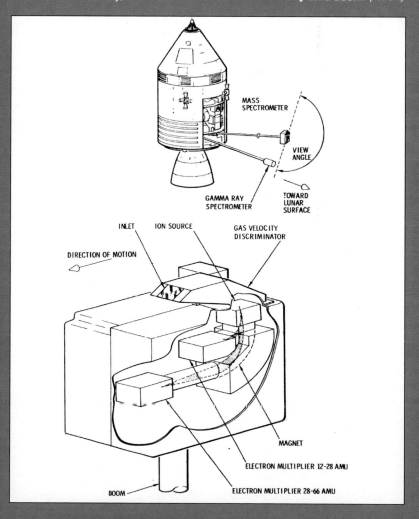

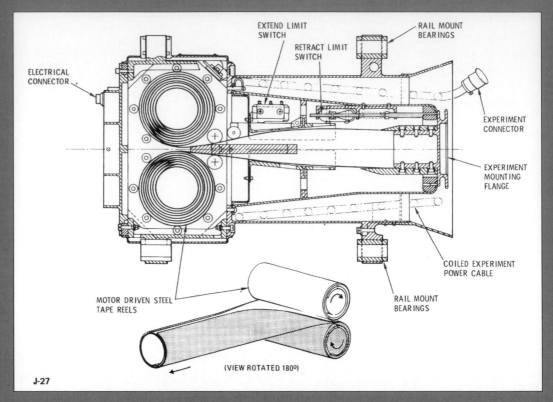

comprising a scanning unit, a thermistor bolometer and the processing electronics. A folded cassegrain telescope with a rotating mirror which swept through 162° cross-track was installed in the scanning unit. Thermal energy reflected from between 7ft (2m) and 200ft (60m) above the surface, reflected off the scanning mirror, was focused on to the thermistor bolometer. The output of this was processed by the electronics package which split the temperature readings into three thermal channels for transmission to Earth as telemetry. During operation from orbit the longitudinal axis of the CSM was aligned to the flight path when this experiment was operating and would be set to run around both day and night sides, mostly while the crew was asleep.

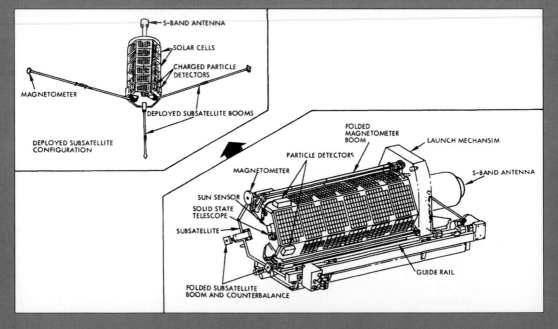

RIGHT The SIM bay configuration for Apollo 17, which differed from that of the two previous J-series flights, as shown here. *(NASA)*

The lunar sounder carried on the J-3 mission provided an electromagnetic impulse beamed to the lunar surface in the high-frequency (HF) and very-high-frequency (VHF) bands to provide recorded data for development of a geologic model of the lunar surface down to the depth of 4,280ft (1,300m). As well as helping acquire data contributing toward an understanding of the stratigraphic, structural, dynamic and topographic measurements, the lunar sounder would also measure ambient noise levels around the Moon at 5, 15 and 150mHz. It would also measure the occultation of the Moon by the electromagnetic waves generated at the lunar surface by an electrical properties experiment transmitter.

The lunar sounder consisted of three major components: the coherent synthetic aperture radar, the optical recorder and three antennas. The two HF dipole antennas, each 24.16ft (7.36m) in length, were deployed outwards from opposing sides of the Service Module and created a span of 80ft (24.4m) across. The VHF Yagi antenna was automatically deployed after the spacecraft separated from the SLA following Trans-Lunar Injection. It had a length of 8.9ft (2.7m). All the operating controls for the lunar sounder were in the Command Module, including provision for jettisoning the three antennas in emergencies. When operated, an electromagnetic pulse would be transmitted toward the Moon and the return signal recorded on film by the optical record device. During HF receive-only operation the other SIM bay instruments would be powered down. Along with the two film cassettes, lunar sounder data gathered on tapes were held in cassettes retrieved by the Command Module Pilot on the return trip to Earth.

RIGHT The general configuration of the lunar sounder and associated antennas for the radar propagation equipment designed to penetrate the lunar surface for its electrical properties. *(NASA)*

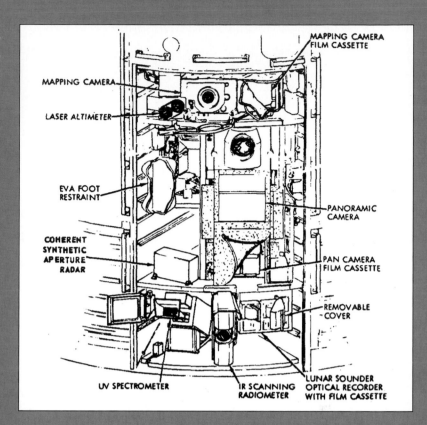

MAPPING CAMERA FILM CASSETTE
MAPPING CAMERA
LASER ALTIMETER
EVA FOOT RESTRAINT
COHERENT SYNTHETIC APERTURE RADAR
PANORAMIC CAMERA
PAN CAMERA FILM CASSETTE
REMOVABLE COVER
UV SPECTROMETER
IR SCANNING RADIOMETER
LUNAR SOUNDER OPTICAL RECORDER WITH FILM CASSETTE

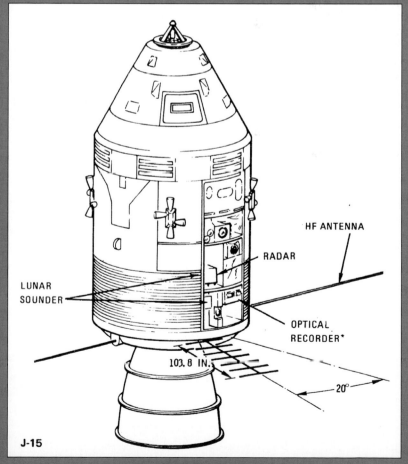

HF ANTENNA
RADAR
LUNAR SOUNDER
OPTICAL RECORDER
103.8 IN.
20°
J-15

ABOVE Cassettes containing film from the Panoramic and Mapping Cameras would be retrieved on the return flight to Earth by the Command Module pilot performing a deep-space EVA, as depicted here in this artist's illustration. (NASA)

A major upgrade to the performance duration of the J-series was the addition of a third cryogenic hydrogen tank installed in sector No 1 as mentioned earlier and this helped provide additional demands for electrical power on the longer flights. The addition of this tank was integral to all the J-series flights and the isolation valve between oxygen tank 2 and 3 was moved from sector No 4 to the forward bulkhead of the Service Module so as to reduce its vulnerability in the event of a catastrophic

failure such as that which happened to Apollo 13. The additional tank held a minimum of 28lb (12.7kg) of reactant, the same as that in each of tanks 1 and 2, and it was mounted in an aluminium sheet metal skirt on a shelf immediately above the SIM and 35in (89cm) below the Service Module's forward bulkhead.

The third oxygen tank was first carried aboard Apollo 14 and was suspended in another skirt below and to the rear of the same shelf. Its recessed position allowed maximum use of this sector space below the cryogenic shelf for the SIM. It was identical to the other two tanks carried as standard in sector No 4 and carried a minimum of 320lb (145kg) of oxygen. Both hydrogen and oxygen tanks in sector No 1 were removable as a single subassembly. Hydrogen tanks 1 and 2 were pressurised by heaters and fans and monitored by a caution and warning light on the main display console but tank 3 was pressurised by fans alone. The heating process was gradual and did not require monitoring. The maximum pressure rise rate was 10.2lb/in² (70.3kPa) per hour, which was slow enough for the maximum vent time of 1.27hr to preclude venting during traversing the far side of the Moon. The available tank 3 flow rate was 0.10–0.24lb (0.689–1.65kPa) per hour. With this improvement in supply and operation, the Apollo spacecraft could return safely to Earth from any part of the mission and have sufficient cryogenics to spare before getting back home.

Given the unique role in conducting the only deep-space EVAs in the history of the space programme for a long time to come, the three Command Module Pilots of the J-series flights were in an exclusive club of their own. The EVA was expected to last up to one hour in two or three sorties back across from the side hatch in the Command Module and along handrails to the aft end of the SIM bay to retrieve cassettes. The EVAs were scheduled for the trip back home so as to avoid being compromised by successive periods of night and day when orbiting the Moon. Nevertheless, this EVA required depressurisation of the Command Module, the first time that had been done since Apollo 9 when Dave Scott photographed Schweickart emerging from the Lunar Module.

To prepare for the EVA, the Apollo spacecraft was first suspended from its usual PTC

BELOW This graphic diagram from an astronaut training manual shows the relative position of hand holds and hand rails together with special shoe retainers by which the crewmember could stabilise himself. (NASA)

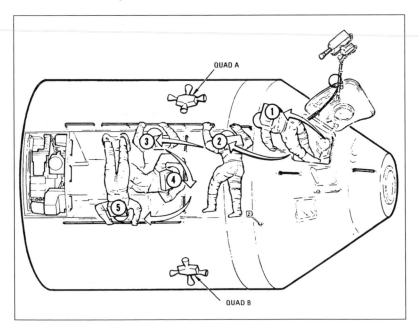

QUAD A

QUAD B

'barbecue' roll mode and stabilised in attitude so as to align the Sun at an appropriate angle for conducting the spacewalk. After attaching a belt to his space suit, the CMP would connect a short tether with a specially strengthened umbilical containing an oxygen hose together with communications lines and biomedical wires. One end of the umbilical was connected to the astronaut's suit, and held from pulling at the suit by connectors attached to a short tether, one end splitting with the tether hooked to the Command Module and the other to a special panel attached to the inside of the spacecraft just below the hatch. The other two crewmembers would hook up their umbilicals to the environmental control system inside the spacecraft.

Before leaving the Command Module, the CMP would add a 30min supply from an oxygen purge system which consisted of two oxygen bottles in a rectangular container which rested on the astronaut's back and was held there by three straps, the same unit as that carried by Moonwalkers above their PLSS units. This would keep him alive if the umbilical failed.

On exiting the Command Module the astronaut would install a campole (camera pole) on a base mounted on the inside face of the hatch, to which he attached a TV and 16mm camera to record the EVA. He then makes his way down the side of the SIM, turns himself around and locks his boots into foot restraints, a period during which the Lunar Module Pilot half exits the Command Module and plays out the umbilical as required. The umbilical was 24.3ft (7.4m) long with a bright orange, 6in (15.24cm) section in the middle as an indicator to stop playing it out any further as the sight of the CMP was partially obstructed around the side of the Service Module.

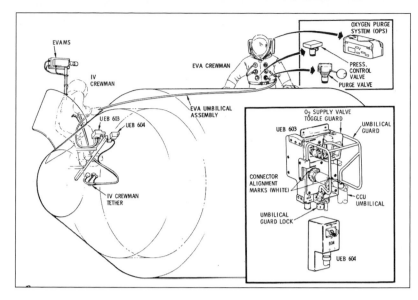

ABOVE The unique arrangement for conducting an EVA to the SIM bay is depicted here showing the special hoses connecting to the environmental control system for supplying oxygen and for carrying communication lines. (NASA)

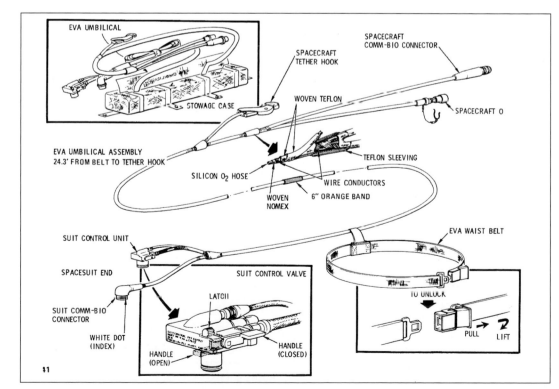

LEFT The special hoses, electrical lines, connectors and tethers to support the EVA were unique to the three J-series missions. In addition, and not shown here, the switch panels located near the Apollo side hatch were protected by special guards placed over them to prevent an astronaut inadvertently activating them. (NASA)

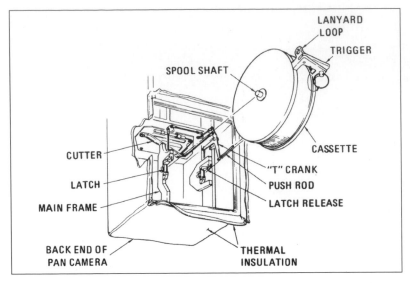

The film cassettes diagram labels: LANYARD LOOP, TRIGGER, SPOOL SHAFT, CASSETTE, "T" CRANK, PUSH ROD, LATCH RELEASE, CUTTER, LATCH, MAIN FRAME, BACK END OF PAN CAMERA, THERMAL INSULATION

ABOVE **The film cassettes were contained in clip-on frames which could easily be disconnected by the crewmembers and returned for stowage in the Command Module.** (NASA)

The CMP first removed the cassette from the mapping camera at the forward end of the SIM bay, the smaller of the two cassettes and weighing about 23lb (10.4kg), which he would attach to his suit by a short tether connected to his wrist. This cassette was then conveyed back to the Command Module and handed, via the LMP, to the Commander inside. The CMP then retrieves the much bulker panoramic mapping camera cassette weighing about 85lb (38.5kg) and returns with that too.

The campole consisted of a 26in (66cm)

titanium tube with a mounting plate for a TV camera and the 16mm camera, the complete assembly being 34in (86cm) in total length. The four handrails on the outside of the Command Module which were installed for access to the Lunar Module should the tunnel be blocked (when the LM is attached) and the three around the docking area would not be used. Each handle was 12in (30.5cm) long and consisted of an aluminium tube with end supports bolted into fibreglass inserts in the heat shield. Newly installed handles along the side of the SIM bay were used for the astronaut to make his way, hand-over-hand, from the hatch to the back of the Service Module. There were also seven handholds inside the SIM, varying in length from 10.5in (26.7cm) to 82.25in (209.5cm).

Extended Lunar Module

The application of the SIM instruments and the cassette retrieval capabilities exploited the scientific value of having the CSM spend longer in lunar orbit but the extension to six days for expanded surveys was predicated on the availability of an Extended Lunar Module available for surface stays of up to three days during which three full working sessions would be completed outside on the Moon.

Like the expanded capabilities of the Apollo spacecraft, work on the Extended Lunar Module really emerged from AAP studies in the mid-1960s. And while, unlike Apollo, the basic LM was a product of the 1962 decision to adopt Lunar Orbit Rendezvous (LOR), that did little to restrain Grumman from working on developed versions with greater capabilities. Realising the potential, NASA invested in additional work on the J-series LM to increase the duration of its operational capability, to expand the science potential and to achieve a greater degree of flexibility.

Most of the modifications were seen on the Descent Stage, with substantial rearrangement of the four quad sectors within the structural

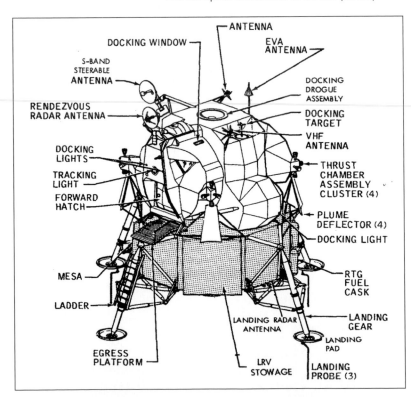

Diagram labels: DOCKING WINDOW, ANTENNA, EVA ANTENNA, S-BAND STEERABLE ANTENNA, DOCKING DROGUE ASSEMBLY, RENDEZVOUS RADAR ANTENNA, DOCKING TARGET, VHF ANTENNA, DOCKING LIGHTS, TRACKING LIGHT, FORWARD HATCH, THRUST CHAMBER ASSEMBLY CLUSTER (4), PLUME DEFLECTOR (4), DOCKING LIGHT, MESA, LADDER, RTG FUEL CASK, EGRESS PLATFORM, LANDING RADAR ANTENNA, LRV STOWAGE, LANDING GEAR, LANDING PAD, LANDING PROBE (3)

LEFT **The Extended Lunar Module had several modifications as described in the text, most notable being the Lunar Roving Vehicle in quad 1, to the left of the forward leg and ladder.** (NASA)

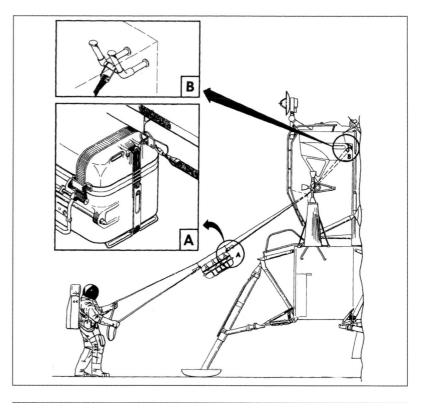

RIGHT Standard on all missions from Apollo 11 but tasked with more use on the J-series flights, the Lunar Equipment Conveyor consisted of a line hooked to the ceiling inside the Lunar Module along which a crewmember could pass equipment up to the porch on top of the ladder. *(NASA)*

frame of the box-structure itself. This provided an opportunity to increase propellant provided for the DPS engine, the electrical power available and the water quantity for crew use and cooling purposes. To vacate those spaces, two 415amp/hr batteries (upgraded from 400amp/hr) from quad 1 and two from quad 2 were relocated to the back of the Descent Stage to join a new fifth battery of the same improved type, all mounted in parallel and providing a total of 2,075amp/hr. Quad 1 was now dedicated to supporting the folded Lunar Roving Vehicle (LRV) which would greatly increase the mobility and traverse range of the crew while relieving them of walking across difficult terrain carrying large quantities of equipment.

Quads 2 and 3 remained unchanged (but with two pallet assemblies in quad 3) but whereas quad 4 housed the MESA, itself a deployable mounting for the surface TV camera, sample containers, antenna cables and basic tools, this was enlarged to carry an additional sample return container, tool pallet, batteries for the PLSS and a cosmic ray detector, all relocated to the exterior of the quad itself. This allowed the interior of the quad to support an additional 111lb (50.3kg) water tank in addition to the existing 266lb (120.6kg) tank in quad 2, a waste container, an additional gaseous oxygen tank to expand total capability to 85lb (38.5kg), and the erectable S-Band antenna (not carried on Apollo 15) relocated from an external mounting on the exterior of the quad.

The additional gaseous oxygen quantity allowed up to six charges of the PLSS backpacks at 1,410lb/in² (9,722kPa) versus six at 900lb/in² (6,205kPa). In addition, urine capacity was increased to 1,200cc/man/day and 100cc/

RIGHT The A7L-B lunar suit was a considerable improvement over the original A7L and key features are identified here. *(NASA)*

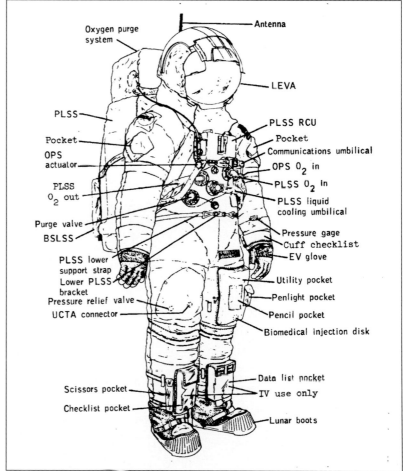

Oxygen purge system

Antenna

LEVA

PLSS

PLSS RCU

Pocket

Pocket
OPS actuator

Communications umbilical

OPS O₂ in

PLSS O₂ out

PLSS O₂ in

PLSS liquid cooling umbilical

Purge valve

BSLSS

Pressure gage

Cuff checklist

EV glove

PLSS lower support strap

Lower PLSS bracket

Utility pocket

Penlight pocket

Pressure relief valve

Pencil pocket

UCTA connector

Biomedical injection disk

Data list pocket

IV use only

Scissors pocket

Checklist pocket

Lunar boots

man/day of PLSS condensate. The two ALSEP pallets were located in quad 3 together with an LRV pallet as well as the Laser Ranging Retro Reflector and some tools which would be relocated to the LRV after deployment. The RTG cask containing the plutonium fuel for providing electrical power to the ALSEP instruments remained at the side of quad 2.

Changes to the Lunar Module Descent Engine (LMDE) expansion skirt were made, increasing its length by 10in (25.4cm) and changing the expansion ratio from 47:1 to 54:1. The 'astroquartz' cone liner from France reduced erosion by up to 50% and increased specific impulse (a measure of efficiency in combustion) by 1.5lb-sec/lb. It also boasted a new, lightweight ablative material providing a net saving of 13lb (5.9kg) in overall weight. The silica combustion chamber liner was also changed to quartz, reducing erosion rates. The four propellant tanks carried within the outer box structures that support the landing gear assemblies were each increased in height by 3.4in (8.6cm), which increased the total propellant load by 6.3% to provide an additional 1,150lb (521.6kg) of fuel and oxidiser.

J-series astronauts also had improved space suits, known as the A7L-B of which two versions were available. The EVA A7L-B was for intravehicular use and for the deep-space EVA which the CMP would perform on the way back from the Moon to access the SIM film cassettes; the Moonwalkers, however, would use a different suit known as the Extravehicular Mobility Unit (EMU), essentially the same but with additional modifications providing added protection and an improved helmet and visor assembly. The EVA suit for the CMP was essentially the same as that worn by the CMP on Apollo 14 but with appropriate fittings for operating from the Command Module.

Both types had a new waist joint that

RIGHT Buzz Aldrin wears the standard A7L suit, without modified visor assembly and without some subtle but important changes which had already been made for Apollos 12 and 14. *(NASA)*

FAR RIGHT Eugene Cernan wears the A7L-B suit with improved design, greater flexibility and being less stressful for the wearer. Never a comfortable garment, the new suit was considered a development in the right direction. *(NASA)*

allowed greater mobility while the suit was pressurised, allowing a greater degree of flexibility for stooping and manoeuvring for attending to instruments and equipment on the lunar surface. There was also an added neck convolute and the zipper was moved from the crotch to assist with leg action and there was provision for reduced shoulder restrictions in articulating the arms. A new full bladder/convolute was provided to help protect against abrasions with an improved zipper design, better thermal blanket and an increased supply of drinking water.

Responding to a lot of feedback from the six Moonwalkers thus far, the helmet had a completely new and greatly improved Lunar Extravehicular Visor Assembly (LEVA) which was mounted to the helmet separately and attached with fittings at each side. The LEVA was not integral to the helmet but an added accessory vital for controlling the amount of Sunlight falling

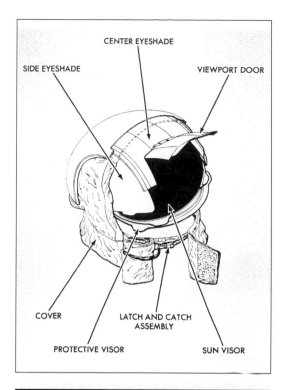

LEFT The modified visor assemblies on the helmet cover helped ease a problem with sunlight on the surface, harsh at best. *(NASA)*

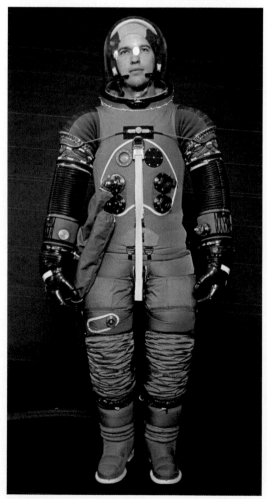

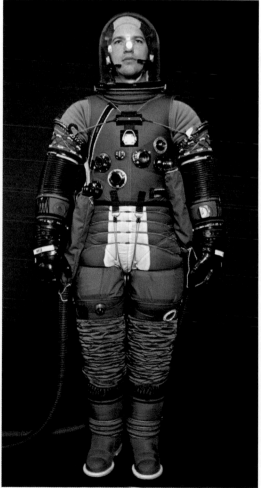

FAR LEFT Seen here without outer layers, the original A7L suit which was satisfactory for a few hours on the Moon but inadequate for extended sessions over three working days. *(NASA)*

LEFT The much improved A7L-B suit over which a new and improved assembly of outer layers would constitute the definitive Apollo Extravehicular Mobility Unit (EMU). *(NASA)*

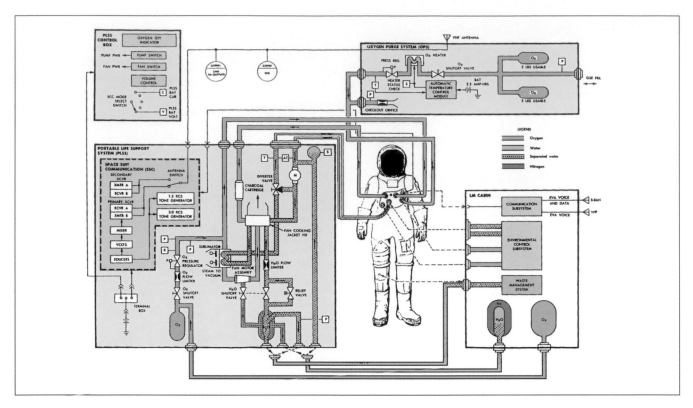

ABOVE The complex interface between the EMU and the Personal Life Support System is displayed in this graphic 'breadboard' layout of interfaces necessary for keeping the occupant alive on the Moon. At left is the PLSS system with the chest-mounted controls box above and the Oxygen Purge System – an emergency pack in the event of a failure in the PLSS – at top right. The interfaces with the LM are at extreme right. *(Via David Baker)*

BELOW The PLSS-2 design took account of the need to significantly improve the operability of the backpack both in quantity of consumables and in the flexibility of controls and options for cooling. *(NASA)*

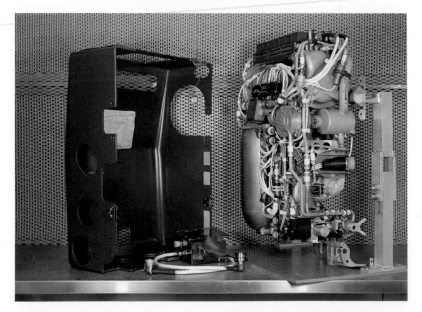

on different facial angles. This was a crucially important part of efficient working on the Moon because the lack of atmosphere produces harsh and unshaded light of much greater intensity than that experienced on Earth.

The new LEVA was attached to clamps at the base of the helmet and provided protection against solar heat, particles, solar glare, ultraviolet rays and accidental damage to the helmet. It consisted of a polycarbonate shell, cover, hinge assemblies, three eyeshades and two adjustable visors for protection from sunlight. The protective visor provided impact, infrared and ultraviolet ray protection while the Sun visor had a gold coating which protected against light and reduced the tendency of the helmet to heat up. Three eyeshades were located, one each side, and one in the centre which reduced glare by preventing light entering at the sides and from overhead. Inside the LM the LEVA units were stowed in the helmet bags and secured to the ascent engine cover.

The PLSS was also upgraded to the PLSS-7 design to support the extended EVAs of around seven hours each. The oxygen pressure was raised from 1,020lb/in² (7,033kPa) absolute to 1,430lb/in² (9,860kPa) absolute,

RIGHT The environmental control system in the
Lunar Module was the hub from which the PLSS
units would be recharged with oxygen and the
carbon dioxide scrubbers replaced for each of
the three EVAs. *(NASA)*

feedwater quantity was increased from 8.5lb
(3.85kg) to 11.5lb (5.2kg), battery capacity was
increased from 279Watt-hr to 390Watt-hr and
the quantity of lithium hydroxide in the CO2
removal system was raised from 3lb (1.36kg)
to 3.12lb (1.41kg). There was no change to the
Oxygen Purge System (OPS) carried on top
of the backpack for emergency purposes but
the 'buddy' system, the BLSS, was an integral
part of contingency equipment, first utilised on
Apollo 14.

Lunar Roving Vehicle

The single biggest operational advantage
with the J-series flights was the Lunar
Roving Vehicle (LRV) and the greater range
which it afforded for planned EVAs and
excursions across the surface. The full
development history, operational details and
technical story of that remarkable vehicle is told
in another Haynes Workshop Manual so only
cursory details are necessary here.

The decision to develop the LRV emerged
from a hotly contested debate between those
who wanted to use a personalised rocket
pack adapted for a one- or two-man Lunar
Flying Vehicle (which itself is to be described
fully in another Haynes title on 'Lost Missions')
or a four-wheeled electric, open vehicle for
transporting astronauts and equipment across
the surface. The Marshall Space Flight Center
backed the car as well as the flying vehicle but
from the outset the latter was disadvantaged
in concept by its inability to simultaneously
carry astronauts as well as large quantities of
equipment, instruments and an increasingly
heavy rock load!

The LFV was championed by the Manned
Spacecraft Center, if only because flying
machines were their forte. The LFV had
emerged from work already conducted by
Textron-Bell Aerospace Company while von
Braun at MSFC was personally backing the car

– he had already told a cynical astronaut Gene
Cernan that he would be driving around on the
Moon within a decade; he was, as Commander
of Apollo 17. But the two competing concepts
crashed into headlong and heated dispute at
a Santa Cruz summer conference on lunar
exploration in August 1967. Work on both
concepts emerged from AAP studies that
supported extended stays on the lunar surface
and the use of a two-launch expedition: one

BELOW Led by Gene
Shoemaker, astronauts
get a geology lesson
in how to recognise
different rock types and
how to discriminate
in selecting samples,
a priority for the
expanded field trips on
the Moon. *(NASA)*

ABOVE The Lunar
Roving Vehicle is
represented here
by a 1g trainer,
designed to support
two astronauts on
extended traverses
of the lunar surface.
(NASA)

with an unmanned Lunar Module providing a habitation module and pre-positioned resources and another delivering the Moonwalkers. But even by 1967 there was little hope of such an ambitious extension to the basic Apollo goal.

Santa Cruz failed to resolve the matter but the issue gathered momentum at a managerial level. With the first landing imminent and the Lunar Module having been proven in manned flight with Apollo 9, in April 1969 the Lunar and Planetary Missions Board recommended an improved space suit, a better backpack and some form of mobility system for extending the range of the astronaut's reach. A vehicle held a lot of advantages – and unlike an LFV it could not crash – and the ability to move the Moonwalkers quickly between sites would make better use of the increased EVA time by reducing the time taken to get between sample sites. Moreover, it would extend the range of opportunities open to the geologists, who could not begin to take a much more active and disruptive role in following traverses on TV and in making direct recommendations to the astronauts about certain actions whenever they stopped to conduct a documented survey.

With time running out and the matter coming to a head, George Mueller stepped in and made the decision to go with an electric wheeled vehicle, with safety arguably the clinching part of the review but with weight-carrying and simplicity of use a close second. The Senior Management Council was convened in May 1969 and approval was given for Sam Phillips to proceed with development. But von Braun knew that it would need a sharp-brained manager and with MSFC charged with the project, he chose former programme manager of the F-1 rocket engine, Saverio Morea, to head up the fast-tracked development effort.

Three companies competed to build the LRV, including Bendix, Grumman and a team consisting of Boeing and General Motors, the latter receiving a contract in November 1969 after headquarters reviewed the final selection which had also included Bendix. Most thought Bendix would produce the winning bid because they had done most of the development work on Moon vehicles, at a time when it was thought that lunar surface bases would be a natural follow-on to Apollo. But overall, Boeing had the better proposal and, at $19.7million, were the cheapest on paper despite the final cost rising to $37million!

By the end of the year the design was getting complex and a meeting in January 1970 cut back on the specification and freed the hand of the design team to come up with a vehicle which, while simplified in its technical requirements, was easier to design, develop, build and deploy. Some rudimentary navigation device was considered essential, to give the astronauts an ability to know which direction they were heading and the direction in which they would have to travel to get back to the LM. And it had to be accurate. With the LRV, the Moonriders would drive out of sight of their only means of survival and if anything went wrong, and the vehicle broke down, they would have to walk back. This latter requirement alone would constrain the distances they could travel, a factor driven in to planning routes, locations and timelines.

But these were not the only difficulties. Some came from the astronauts themselves. Respected for their pioneering experience, Armstrong and Aldrin came back convinced that there was no need for a vehicle, that moving around on the Moon was easy and that even at a leisurely lope, two men could gain on a specified station stop in less time than it would take to extract, deploy, set up and operate a vehicle to take them across the same distance. And they made these views known, much to the disturbance of the planning teams. But Apollo 11 had not provided the schedule, or the activity, which would demonstrate the limits to which humans could physically move across the lunar surface. When Shepard and Mitchell demonstrated such limitations as they struggled to carry their Modularised Equipment Transporter up the steep slope of Cone Crater it proved the point: astronauts could not conduct useful work without some form of transport. And on J-series missions they would be carrying a lot more equipment.

The first LRV arrived at the Kennedy Space Center on 15 March 1971, just over four months prior to the first J-mission, Apollo 15, with weight having grown to an overall 460lb (208kg), allowable weight with payload of 1,080lb (490kg) including the two suited astronauts, 100lb (45kg) of communications equipment, 120lb (54.4kg)

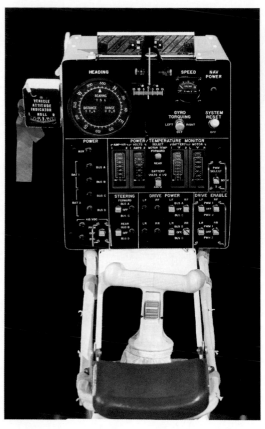

of scientific equipment and instruments including photographic equipment and 60lb (27kg) of lunar rocks. It also had a reduction in accumulated range on three EVAs of 40ml (64km), down from a planned 72ml (115km). That latter restriction was not because of an absence of capability but because mission planners wanted the crew to stay within 6ml (9.6km) of the Lunar Module for that walk-back criteria in the event of a breakdown, at the most distant point at the end of an EVA.

But viewers on Earth disillusioned by the absence of any surface TV on Apollo 12 and the inability to see Shepard and Mitchell struggling up to Cone Crater on Apollo 14 (the camera had been left just looking at the vacated LM), were about to get an extraordinary display of real-time lunar exploration. For not only would the LRV carry its own communication system and colour TV with it all along the expeditionary routes, at the end of the mission it would be positioned to show the lift-off from the Moon and track the Ascent Stage as it disappeared into the blackness of space.

The LRV had a total length of 10.16ft (3.01m), a 6ft (1.8m) tread, a 7.5ft (2.3m) wheelbase and a height of 4.3ft (1.3m). Total motive power from four independent motors, one to each wheel, was approximately 1hp (0.745kW) with a top speed of about 8mph (12.8kph). With Ackermann steering it had maximum manoeuvrability and each steering system could be isolated, or the other used if one failed.

Because of the unique nature of its job and the environment in which it was designed to work, the mobility system was designed with technology more akin to the space programme it served than the automotive industry from which it inherited its configuration. Fabricated from aluminium, its chassis was separated into forward, centre and rear sections, integrally separated for stowage but locked when deployed. The forward section housed the batteries which powered the wheels as well as the systems for navigation and steering and the communications equipment. The LRV became an independent broadcasting platform with a dedicated antenna direct to Earth, severing the link through the Lunar Module on which the first three missions relied.

The centre section was essentially the crew station with two foldable seats, controls and displays and stiffened to support the full weight of two suited astronauts in 1/6th gravity. Fabricated from beaded 2024 aluminium panels, it also served to rigidise the front and rear sections. The latter consisted of a platform

on which the science payloads and tools would be stowed together with portable instruments and equipment carried along for use at selected station stops. Two adjustable footrests, folded flat to allow chassis folding, were deployed by the crew and could be used in the up or down position. Seat belts fabricated from nylon webbing were provided for each occupant, with Velcro strips holding the two sections together. The crew were protected from fine dust by glass-fibre guards, the aft section of each only deployed to its lower position when the LRV was on the lunar surface.

The wheels were formed of woven mesh of zinc-coated piano wire on which titanium treads were riveted in a chevron pattern, attached to a spun aluminium hub and bump stop. Considerable research had gone into the appropriate sort of wheel for use on the airless Moon and much work had been conducted into the behaviour of various materials, from rubber tyres to solid wheels and from steel-rimmed, open mesh wheels to leg-like contraptions attached to a circular drum-shaped hub. The design of the weave was modelled on replicas of looms found in excavated remains of workshops in Ancient Egypt.

A single LRV wheel, 32in (81.3cm) in diameter and 9in (22.8cm) wide weighed 12lb (5.4kg) and was designed to outlast a total driving distance of 112ml (180km). The absorption value of energy and loading took account of the 1/6th lunar gravity and was carefully integrated with the traction and suspension system. A special 1g trainer was used for the astronauts to practise with since no flight-rated LRV would be able to support itself on Earth, weight and the wheel structure being designed exclusively for the reduced gravity of the Moon.

Traction was designed for safety-first considerations and with an integral mechanical brake connected by cable to the hand controller; moving it rearward energised the electric drive motors and pushing it forward activated hinged brake shoes working against a cylindrical drum. Full rearward movement of the controller placed locks on each wheel to engage a parking brake. To disengage, the controller was moved to the left position, releasing the brake and returning the controller to neutral.

BELOW The general configuration and dimensions of the Lunar Roving Vehicle designed to fold into three separate chassis sections for packing in quad 1 of the Lunar Module Descent Stage. (Boeing)

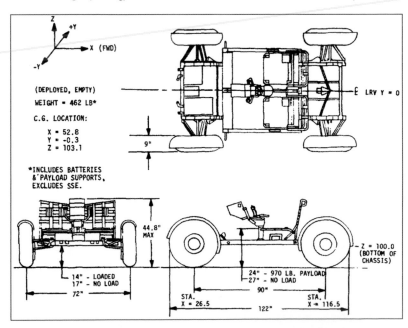

The suspension included a pair of parallel arms for each wheel mounted on torsion bars connected to each traction drive with a shock absorber as an integral part of each suspension system. Under maximum loading there was a ground clearance of 14in (35.5cm) which was sufficient for crossing undulating surfaces littered with small obstructions and rocks and shallow crater rims. Total steering radius was identical to the length of the LRV with a small vane added between the chassis and the steering arms to permit extreme steering angles. A further safety provision was the ability to disengage a fouled or obstructed wheel and continue on with the remainder operating.

Driver controls include a T-handle on the central display console midway between the two seats through which all operations including forward, reverse, steering, speed and braking could be controlled. Selection of forward or reverse was managed via a button with associated electronics controlling power, drive and elements of the navigation system, which was a crucial part of the safety element in operating the vehicle at increasing distances from the LM – limited to 6ml (10km) by the 'walk-back' timeline outlined earlier.

The Moon has no magnetic field with an aligned polarity like Earth so a unique form of navigation was required for the crew to know how to navigate around on the surface. Employing a system of dead reckoning, the LRV was equipped with a directional gyroscope, odometers on each wheel and a small, solid-state computer for determining directional vectors, distance traversed and speed. The navigation system required the displays to be set before moving off, a compass on the central display console being set to lunar north at the start of the traverse. The compass dial itself showed heading in degrees around the circular display in segments of 30°, with three digital displays within the inner section of the compass showing heading, distance travelled and range back by direct route to the Lunar Module. The computer calculated the bearing the LRV must travel to return to the LM so that the astronauts always had that information to get them back into visual range of their spacecraft.

Power loss to the navigation system would lock up the bearing indicator to preserve the

reading so that even if the LRV was disabled the two men could head off in the correct direction on foot. The distance indicator took its readings from the signal processing unit which received nine odometer magnetic pulses for each wheel revolution. For distance a sample measurement was taken from the third fastest wheel only and read out on the display in increments of 0.1km (328ft). An interesting aspect of lunar science in the Apollo programme was that the flight dynamics of the spacecraft were in imperial units while all units relating to activity on the surface were in metric. Like the bearing, range was preserved in the event of a total power failure.

Centrally mounted on the upper portion of the displays was the most critical element in the navigation subsystem. Called the Sun-shadow device, it consisted of a hinged plate with a hole drilled through the centre to reveal a horizontal scale. When the plate was lifted to the horizontal position, with the LRV facing directly down-Sun, a shadow would be cast on the scale beneath. With a reading from the attitude indicator the crew would inform Mission Control of the values on the shadow scale. This device enabled the crew to determine their heading with respect to lunar north at any time. The graduated scale

ABOVE This verification of deployment procedures shows the tight confines into which the vehicle was installed for flight to the Moon's surface. *(NASA)*

ABOVE **This view shows how the LRV wheels folded in upon the centre section of the chassis, itself folded upon by fore and aft sections.** *(NASA)*

automatically lower to preserve the semi-passive thermal control and to prevent the build-up of dust on the reflective surfaces.

A major advantage for the J-series programme objective and one which would greatly benefit the acquisition of an expanded science data set, the LCRU relieved the Lunar Module of responsibility for relaying communications to and from Earth as it had on the two H-series landings. Acting as a portable relay station for data, voice and television, it literally took Mission Control along with the astronauts, providing scientists with the opportunity to amend or change specified activity at selected station stops designated for attention pre-flight. This also engaged the public in a much more vivid and exciting experience and restored to public relations officials the opportunity to give viewers continuous, unbroken coverage all the way out and all the way back.

Although transported on the LRV, the LCRU was completely independent and incorporated its own power supply and folding S-Band antenna, means by which Houston could control the TV system while the crew were working close by as well as affording the opportunity to transmit a televised view of the lunar lift-off. At each traverse stop a crewmember would boresight the parabolic high-gain antenna with Earth and VHF signals from each astronaut's PLSS transceivers would be converted to S-Band and transmitted to Earth, the converse being the case for signals flowing from Earth to the LCRU and thence to each crewmember.

With a size of 22in x 16in x 6in (55.9cm x 40.6cm x 15.2cm) and weighing a mere 9.2lb (4.1kg) by lunar weight, it could be hand carried or stowed on the LRV. The Ground Commanded Television Assembly (GCTA) was integral to the LCRU and did what it said, supporting an RCA colour camera for lunar surface use, being utilised from the MESA during initial astronaut descent down the ladder, installed on a tripod and connected to the LM by a 100ft (30m) cable, or installed on the LRV and connected to the LCRU. The camera itself could be moved by an astronaut or operated from Mission Control including pan, zoom, tilt, aperture setting and light control. The GCTA was capable of elevating the camera upward 85°, downward 45° and 340° in pan between stops; motion in pan and elevation was 3°/sec.

spanned +/-15° and could be used up to a Sun angle of 75° from the horizontal.

Power for the LRV was provided by two 36-volt, non-rechargeable batteries, of plexiglass construction using silver-zinc plates in a potassium hydroxide electrolyte with a magnesium outer casing. Each battery had a 121amp/hr capacity and provided power for the mobility subsystem, gyroscopic unit signal processing unit, display console and the navigation equipment. A 36-volt utility outlet provided a 150W current to a point on the front of the LRV for powering the Lunar Communications Relay Unit (LCRU) and the TV camera.

Battery temperatures were maintained between 40°–125°F (4.4°–51.6°C) by a passive thermal control system with heat transmitted to the drive control electronics, signal processing unit and directional gyro unit and conducted to the two batteries along aluminium thermal transfer straps. The display and control electronics, the signal processing unit and the two batteries carried fused silica mirrors on their upper surfaces and at the completion of each sortie the fibreglass blankets were raised to allow the heat to escape through radiation. With the batteries cooled to their normal operating temperature of 45°F (7.2°C), the shrouds would

Boeing was originally expecting to build six test units but that increased to seven and within two months of go-ahead they had prepared a static mock-up of the original concept. Throughout December 1969 the Marshall Space Flight Center co-ordinated an evaluation of subsystems layout and crew station design. Astronauts Gerald Carr and Joe Engle played a prominent role in these early development stages of the vehicle.

During the first half of 1970 work progressed on the other units. One of the first requirements of the development phase was the evaluation of stress factors imposed by the mass of the vehicle suspended on the side of the LM. A so-called 'LM-LRV-Glob', a weird, angular contraption duplicating the mass properties of the LRV, was placed in the stowage bay of the test version of the Extended Lunar Module. Much information was obtained on the stresses and strains caused by vibrations and shocks imparted to the bay, leading directly to the programming of computer simulations of a complete mission. The LM/LRV was assembled for the Marshall Space Flight Center and, liaising with the Grumman Aerospace Corporation, verified mating the LRV to the LM and the response of the combined configuration.

The third test unit to be completed was a 1/6th gravity mock-up used to evaluate the deployment mechanism, simulating under Earth conditions the exact weight of the LRV when undergoing deployment on the Moon. Shipped to Boeing's Kent, Washington, facility in August 1970, a vibration unit was subjected to all the rigours of launch, boost and the stresses produced by the LM propulsion systems during flight. Essentially a structural mock-up, the vibration unit used weights to simulate the electrical and crew station elements to duplicate the flight hardware. For two months vibration tests continued, with information obtained through strain gauges and accelerometers applied to refinements in the final design.

Meanwhile, activities at Marshall had centred on crew familiarisation with the evolving design and included the simulated attachment of TV camera, communications antenna, tool carriers and geological equipment to determine their optimum placement. During August, Delco

Electronics prepared a mobility test unit to prove the validity of the upcoming 1g trainer. Powered through umbilical connections from a truck, astronauts Jack Lousma and Gerald Carr developed control procedures and tested the response of the mobility subsystem, wheel and steering subsystems.

In October 1970 Delco completed the long-awaited 1g trainer and this was delivered to the Marshall Space Flight Center on 17 November. Powered by rechargeable nickel-cadmium batteries, the wheels were fitted with inflatable tyres, added strengthening was applied to the chassis and suspension unit and thermal control maintained by fans. However, the trainer had been used to provide a full simulation of the control and stability characteristics of the flight vehicles and was used to rehearse anticipated excursions on the lunar surface.

The last test item, the qualification unit was completed in November 1970 and this was identical in every respect to the flight hardware and was used in simulating the effects of vibration, temperature extremes and the vacuum environment of the Moon. Assembly of the first flight-qualified LRV began in December and was moved from the fabrication facility to begin tests on 4 February 1971. The new Director of the Marshall Space Flight Center, Dr Eberhardt Rees was at the formal handover ceremony on 10 March. Five days later, little more than 16 months after signing the contract, NASA received LRV-1 at MSFC and for the remainder of the time until the launch of Apollo 15 tests were sustained with a separate chassis unit.

BELOW Fully deployed, the rover is open for business as it will be for the first time on Apollo 15. *(NASA)*

J-1 Apollo 15

Launch date: 26 July 1971
Duration: 295hr 11min 53sec

The crew for Apollo 15 was commanded by David R. Scott, veteran of Gemini VIII and Apollo 9, and two astronauts making their first flight, Command Module Pilot Alfred M. Worden and Lunar Module Pilot James B. Irwin. The back-up crew consisted of Richard F. Gordon, Vance D. Brand and scientist-astronaut Harrison H. Schmitt in respective positions; this crew would have cycled back on as the prime crew for Apollo 18, one of the now cancelled flights. None of the Apollo 15 crew would fly again.

The launch of the first J-series missions began on time and on date at 09:34:00 local time, 26 July 1971, watched by several hundred thousand people across the area from Cocoa Beach, north up the coast and inland. Among them was a contingent from the British Interplanetary Society (BIS) including Executive Secretary Len Carter, aerospace engineer and Editor of *Spaceflight* magazine Ken Gatland and renowned space artist David A. Hardy, as well as several well-wishers who had themselves played a not insignificant role in this, the first major scientific expedition to the Moon. Appropriately, at a 'Manned Flight Awareness' reception given by NASA for leading members of the government–industry team responsible

for the mission, NASA Deputy Administrator George M. Low made a generous reference to original plans by the BIS from 1937–39 outlining the design of a lunar lander and mentioned attendees in person.

With a significant upgrade to the payload capability and an improvement in its performance, AS-510 was not the heaviest Saturn V to this date but it was carrying the heaviest mass of combined spacecraft ever lifted from Earth. But weight saving had improved the mass/payload fraction; at ignition, AS-510 weighed 6,494,415lb (2,945,817kg) and at lift-off 6,399,652lb (2,902,833kg), compared with 6,505,548lb (2,950,866kg) and 6,417,871lb (2,911,057kg) respectively for AS-509 launching Apollo 14.

Nevertheless, CSM-110 launched by Apollo 14 weighed 64,388lb (29,206kg) at transposition and docking shortly after departing Earth orbit for the Moon, while for that mission LM-8 weighed in at 33,649lb (15,263kg). But its successor was much heavier.

Incorporating all the upgrades and additional elements, for Apollo 15, CSM-112 weighed 66,885lb (30,339kg) and LM-12 weighed 36,220lb (16,429kg). Together the docked vehicles for Apollo 15 weighed 103,105lb (46,768kg), an increase of 5,118lb (2,321kg), or 5.2% more than its H-series predecessors.

The additional lift potential which allowed these heavier vehicles to fly on the tenth Saturn

RIGHT The crew of Apollo 15. From left Jim Irwin, Al Worden and Dave Scott, together with the mission badge. *(NASA)*

V was helped by more favourable temperature and wind effects encountered by launching in July rather than in a winter month. Also, by a slight reduction in the propellant reserves (based on prior experience with earlier Saturn Vs), there was increased propellant loading for the first TLI opportunity and improvements to the likelihood of high oscillations during powered flight by installation of a helium gas accumulator in the S-II centre engine.

Another change opened a wider launch azimuth range of 80°–100° instead of the usual 72°–96°. Due to its on-time launch, Apollo 15 flew a flight azimuth of 90° with a roll manoeuvre at 12sec to an azimuth of 80°. In terms of energy output, the massive F-1 engines were re-orificed which increased propellant flow rate from 28,814lb/sec (13,070kg/sec) to 29,588lb/sec (13,421kg/sec) and a nominal theoretical thrust per engine of 1.522million lb (6,769.8kN).

Significant weight reduction had been obtained by removing four of the eight retro-rockets in the base of the S-IC first stage, which on prior experience were deemed unnecessary, and all four of the S-II second stage ullage rockets which on previous missions had provided added positive forward motion to settle the cryogenic propellants in the base of respective tanks, also now deemed no longer necessary. Total stage thrust at lift-off was 7.558million lb (33,618kN), a little less than predicted, but the lighter S-IC stage still produced a higher initial acceleration reaching Max-Q – the period of maximum atmospheric pressure on the launch vehicle – five seconds earlier than AS-509 while encountering a record 732lb/ft² (5,047kPa) versus 667lb/ft² (4,599kPa) with the previous flight.

The centre F-1 shutdown occurred at 2min 16sec with outer engine shutdown at 2min 40sec. S-IC cut-off came 9,000ft (2,743m) lower and 200ft/sec (61m) higher than that for the previous flight and with the first stage gone the trajectory was optimised to achieve velocity rather than altitude. At this point the extra weight of the payload, no longer compensated by a very much lighter first stage, contributed a near 9,000lb (4,082kg) deficit for the second stage to propel.

Even with the very much more powerful first stage the heavier upper stages would never

ABOVE The full complement of J-series hardware for Apollo 15, including LRV-1, LM-10 and CSM-112. *(NASA)*

BELOW The Apollo 15 crew showing the disposition of the LRV, the communications antenna for direct link to Earth stations and the subsatellite mock-up. *(NASA)*

have reached the 25,568ft/sec (7,793m/sec) required for the standard 115ml (185km) orbit. Once again the trajectory had to be optimised by targeting orbital insertion 11.5ml (18.5km) lower than usual, which, remarkably, required a velocity only 10ft/sec (3m/sec) greater than AS-509.

However, by deleting four of the five S-IC retro-rockets, the first stage loitered closer to the S-II after separation and was only 20ft (6m) behind the second stage when its five J-2 engines ignited. It had been expected that the

separation distance would have been at least twice that, instead of the 60ft (18m) separation when fitted with eight retro-rockets (as all previous Saturn Vs had been). Being much closer than expected, the plume from the S-II's engines impinged more directly on the forward dome of the S-IC and damaged some of the telemetry equipment in that area.

Moreover, post-flight analysis revealed that if one of the four retro-rockets had failed to fire there was a serious risk of the S-IC shunting into the base of the S-II and its five rocket motors. Had that happened it would undoubtedly have caused destruction of the J-2 expansion skirts (nozzles) causing an in-flight abort and the total loss of the stack; almost certainly, the abort would have triggered the Launch Escape System and the crew would have been saved – but the mission would have been lost. Subsequent Saturn Vs would revert to eight retro-rockets on the first stage to ensure greater deceleration and adequate separation distance.

S-II ignition came at 2min 42sec with centre-engine cut-off at 7min 40sec followed by the outer four J-2 engines at 9min 9sec, some 1.2sec earlier than predicted due to a slightly higher thrust than anticipated. Third stage ignition came four seconds later with shutdown at 11min 44sec, four seconds early due to it too having a slightly higher thrust level than expected, placing the assembled stack in an orbit of 106.5ml (171.3km) x 105.3ml (169.4km), rather than the circular 103.5ml (166.5km) targeted. Following a checkout of systems and a trajectory measurement from Houston, a slight navigational correction was uplinked to the crew prior to TLI which occurred with re-ignition of the S-IVB at 2hr 50min 03sec, a burn lasting 5min 53sec putting projected pericynthion at 160ml (257.4km), rather than the 78.25ml (125.9km) pre-planned.

Apollo 15 was inserted into a non-free return trajectory at TLI, the first time that had been adopted, a flight path that would not put the spacecraft on a path that would automatically swing around the Moon and bring it back to Earth. Readers will recall that the initial Moon missions of Apollos 8, 10 and 11 had employed free-return trajectories whereas Apollos 12, 13 and 14 had flown hybrid trajectories in which

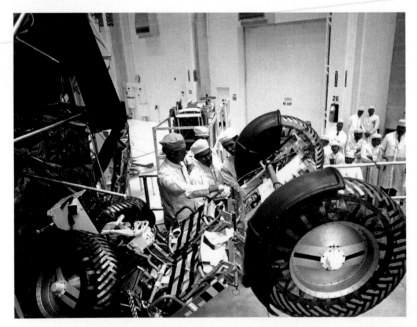

the docked vehicles were switched to a non-free return type only after the Lunar Module had been extracted and the descent engine made available for restoring a return path should the SPS engine be found inoperative.

But the landing site for Apollo 15 was at 26.1°N and that meant it could only be accessed via a non-free return path from the outset. The subtle difference between this type of non-free return path and that of previous missions was that this was the first time the trajectory could only be restored to a free return type if significant RCS thruster ΔV was applied; the earlier free return type would rely on RCS thrusters only for course corrections.

But that presupposed that any correction back to a truly free return type would have to be made before an elapsed time of 50 hours, after which the trajectory would have automatically migrated to a fully non-free return type where even the RCS thrusters would be insufficient to get the crew back to Earth. In this regard it was itself a hybrid derivative of the free return type, only allowable within safety constraints because of the extraordinarily low probability of both SPS and RCS failing on the Service Module in a situation where the LM too was disabled!

As with Apollo 14's S-IVB, the stage would be directed to a lunar impact providing the surface seismometers with shock waves and to achieve this the auxiliary propulsion system in side-pods on the exterior were fired at 5hr

47min 33sec with a second burn at 10hr-even, about 30min later than expected to allow time for tracking to accommodate any flight path changes caused by venting residual propellants in the tanks. The S-IVB would impact the lunar surface at 79hr 24min 42sec near the crater Lalande about 91ml (146km) from the designated spot but sensed by the Apollo 12 and 14 seismometers.

Meanwhile, 27 minutes after TLI the CSM was separated, the SLA ejecting at that time, and transposition and docking was completed

LUNAR ROVER VEHICLE TRAVERSES

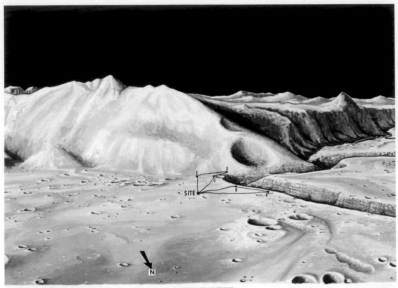

WALKING TRAVERSES

ABOVE Uncertainties regarding the LRV caused operations planners to prepare walk-only traverses for the three EVAs, as indicated here. *(NASA)*

at 3hr 33min 49sec, without any of the problems with the docking probe experienced on Apollo 14. However, almost immediately Scott reported that an 'SPS Thrust' light came on in the Entry Monitor System Delta V displays and telemetry indicated that the solenoid drivers were on. It appeared that a switch or some adjacent wiring was at fault but further resolution was deferred to a mid-course correction burn.

The first opportunity for a course correction adjustment was cancelled but an SPS did take place at 28hr 40min 30sec in which the

RIGHT Dave Scott conducting geology training in Hawaii seven months before the flight, a location patronised along with Iceland by the geologists training astronauts in basic principles and field surveys. *(NASA)*

engine fired for 0.7sec to change velocity by 5.3ft/sec (1.61m/sec) lowering pericynthion to 72.5ml (116.6km), slightly lower than required (see above). That required a second course correction at 71hr 31min 14sec in which another SPS burn, for 0.9sec, trimmed pericynthion to precisely the pre-planned value.

These two SPS firings provided an opportunity to evaluate the spurious readings seen earlier and isolate the fault to an intermittent short. The SPS redundant engine firing path could go through Bank A or Bank B circuits and Mission Control was assured that one of those could be used for all computer-controlled firings while the other bank could still allow manual operation of that engine. While seeming trivial, the determinations ensured mission rules were not contravened and that there was sufficient reassurance to proceed with lunar orbit operations.

Between these two engine burns, Scott and Irwin entered the Lunar Module, named Falcon, at 33hr 56min to check it out, leaving Worden to look after the CSM, named Endeavour after Captain Cook's famous vessel. A 50min TV broadcast revealed a broken cover glass over one of the tape-meter displays but two covers are provided and this was no problem.

On the way to the Moon the crew attached their 16mm data acquisition camera to the sextant in the navigation bay and shot test views of selected stars. Then they used the Lunar Module Pilot's side window (suitably modified with a special quartz pane) to photograph the Earth in ultraviolet with the 105mm Zeiss Sonnar lens attached to the electric Hasselblad camera. This black-and-white photography could be used in studies of planetary atmospheres. The 'flashing light' phenomena also received attention, with the crew spending nearly one hour facing different directions, closing their eyes and reporting the apparent flashes and streaks of light theorised to be a registration of cosmic rays on the optic nerve creating the illusion of light.

A second inspection of Falcon began at 56hr 26min to vacuum up broken glass from the tape-meter display and close up the LM before crossing into the gravitational influence of the Moon at 63hr 55min 20sec. Two more sleep periods and a leaking water spigot occupied the crew during the traditionally quiet translunar coast period. Shortly after the second course correction, at 74hr 06min 47sec the SIM bay door was pyrotechnically jettisoned, exposing the battery of scientific equipment to be used in lunar orbit, the crew watching as it slowly tumbled away into space. This was considered a potentially hazardous event and the crew suited up just in case the explosive charge triggered a pressure leak, of which there was none, the astronauts noticing a solid but dull 'thud'.

The insertion into lunar orbit (LOI) occurred when the SPS engine was ignited at 78hr 31min 46sec, this time for a long 6min 40.7sec punctuated by Bank A being shut down 32sec prior to the planned cut-off time so as to evaluate the performance of Bank B for single-bank burns which may be used in the future. At 195.7ml (315km) x 66.4ml (106.8km), the tracked orbit following acquisition of signal around the right-hand side of the Moon was close to the planned values. As with Apollo 14, the Descent Orbit Insertion (DOI) burn would place the docked vehicles into the elliptical path necessary for Falcon to begin its descent to the lunar surface the following day. The DOI burn with the SPS engine began at 82hr 39min 48sec around the far side of the Moon on the second orbit, lasting 3min 34sec and placing Endeavour and Falcon in a path of 67.3ml (108.3km) x 10.5ml (16.9km).

Following a rest period, the crew were awake by 93hr 35min and preparing for their descent to the plain nestled in a northern bay of the longest mountain range on the Moon. Before separation of the two vehicles it was necessary to trim the elliptical orbit to prevent the mass concentrations of gravitational anomalies within the Moon from wrestling the orbit into a dangerously low pericynthion. Performed with the RCS thrusters on Endeavour, the 30sec burn started at 95hr 56min 45sec, raising the orbit from 67.9ml x 8.2ml (109.2km x 13.2km) to 68.9ml x 11ml (110.8km x 17.7km). This was necessary to re-establish the predicted altitude at PDI so as not to compromise the guidance and navigation and the programmes in the

RIGHT Saturn V AS-510 rolls out from the Vehicle Assembly Building at the Kennedy Space Center on its way to Launch Complex 39A. *(NASA)*

THE HADLEY APENNINE LANDING SITE

The most prominent mountain range that forms a segment of the rugged chain encircling Mare Imbrium is the Apennine Mountains, a segment of the rim created when the basin was formed around 3.9billion years ago. Forming the south-eastern boundary of this impact basin it wanders for nearly 450ml (724km), punctuated in the south by the crater Eratosthenes and coming to an abrupt end north of Mount Hadley on its eastern flank. To the west lies the Palus Putredinus (Marsh of Decay), a bay providing smooth outflow from the lower foothills of the Apennines. To the east of this mountain range lies the Mare Tranquillitatus (Sea of Tranquillity), the two mare surfaces flowing into each other through a gap 50ml (80km) wide. The northern edge of this gap rises to the Caucasus Mountains which themselves curve upward to give way to the Alps bounding Mare Imbrium in the north by the crater Plato.

Thus, the Apennines, the Caucasus and the Alps follow the entire eastern half of the Mare Imbrium, prescribing a spectacular semi-circle of rugged peaks. Interest has long been centred on the formation of Mare Imbrium itself and many theories had been put forward to explain the existence of such areas. It was generally accepted that the giant plain was the result of a major impact that threw ejecta across several hundred miles over the face of the Moon. As with Apollo 14, this fourth lunar landing searched, in part, for an answer to the riddle of this early phase in the evolution of the lunar surface.

With any sizeable impact the material ejected from the deepest part is deposited around the rim. The Apennines are, in all probability, a part of the rim formed when Imbrium itself was torn from the primordial magma. Samples from this area could be expected to predate the Mare surface material. The Fra Mauro site visited by Apollo 14 returned samples from ejecta near to the surface of the Moon when Imbrium was formed. By examining the area close to the rim of Cone Crater the crew were able to gain access to this material, raised to the surface by the impact that created Cone. However, Cone Crater lies some 360ml (579km) to the south of Imbrium and to study the ejecta from the depths of the Moon it was necessary to go to the rim of Imbrium itself.

To satisfy both this and other requirements it was necessary to search out an area within these mountains onto which Apollo 15 could set down. Comparison with data from Fra Mauro and the site on the very rim of the Imbrium ejecta would hopefully enable a reasonably accurate fix to be made on the early formation of surface features. Having established the area, another feature of great scientific interest was Hadley Rille. Following a snaking route for some 60ml (96km), it lies directly south of a much larger complex of gorges known as the Fresnel Rilles, and but for the presence of Hadley Mountain to the east would form a part of the same family. Up to 0.5ml (0.8km) wide and more than 1,000ft (300m) deep, Hadley Rille is only one of several hundred gorges that can be found on the lunar surface.

Their origin is unexplained and selenologists are at a loss to satisfactorily fit them to any one of several evolutionary models, other than to categorise them as collapsed lava tubes, the circuitous route of liquid magma which flowed beneath the surface of a dried lava lake and which cooled allowing the upper part of collapse and leave the rille bed littered with rocks and boulders.

Several alternative landing site choices had been considered for Apollo 15 that would have sampled other lunar rilles. In particular, a site on the western side of the Moon in the Marius Hills was considered. The Marius Hills were believed to be volcanic structures, but their dome shapes suggested to some scientists that they formed from relatively viscous lava, possibly of a different composition from the Mare basalts sampled elsewhere on the Moon. However, it was felt that the Imbrium Basin rim was a more important target than the unusual volcanism in the Marius Hills. Moreover, the Marius Hills are nearly on a straight line drawn through the Apollo 12 and 14 landing sites. This was not a favourable configuration for seismic studies using the passive seismometers deployed on each of these missions. The Hadley Apennine site provided a much more favourable seismometer network geometry.

LEFT An image of the Apollo 15 Hadley Apennine landing site taken during the mission provides a good view of the rille and the adjacent area where the Lunar Module would land. To the left is the vast expanse of the Mare Imbrium. *(NASA)*

Lunar Guidance Computer which would control the descent.

With Gene Kranz taking over in Mission Control, Scott and Irwin entered Falcon a little early, at 97hr 35min and received a 'go' for undocking shortly before passing out of sight on revolution 11 nearly two hours later. When contact was re-established it was learned that the two vehicles had in fact been unsuccessful in separating and that they had been unable to get the docking latches to disengage. A check of the electrical umbilicals in the docking tunnel resolved the problem – a connector was loose – and separation was effected at 100hr 39min 16sec, 25 minutes later than scheduled.

An hour later, at 101hr 38min 59sec the SPS engine fired up yet again for a 2.7sec burn to raise Endeavour's pericynthion, placing it in a more nearly circular orbit of 74.4ml (119.7km) x 61ml (98km). In that path, Apollo, acting as the mothership, would be ready for any abort that might be necessary during the descent. After navigational checks, landmark tracking and final switch and controls configuration, Falcon was ready for its descent to the surface but this time the descent trajectory would follow a path at 25° to the normal horizontal, versus 16° for Apollo 14.

By approaching at a steeper angle the spacecraft had better terrain clearance and by coming down on that path the LPD would be available slightly earlier and with greater visibility over the target area. The impact of this would also be felt on the time Falcon reached High Gate around 8,000ft (2,440m) above the surface which on Apollo 14 had occurred at 2min 58sec prior to planned touchdown and some 25,000ft (7,620m) uprange of the designated touchdown spot. On this mission, High Gate would be 2min 48sec before nominal landing time and less than 16,000ft (4,875m) from the site. Pre-flight planning had Falcon touch down at 12min 02sec after PDI.

Ignition for descent occurred at 104hr 30min 09sec with the usual power profiles as identified for previous flights and the same general sequence of events. At 1min 35sec Mission Control fed the crew with updates to shift the pre-planned landing point 2,800ft (853m) downrange and at 3min 1sec the spacecraft yawed to face up toward space, readying the orientation of the spacecraft for pitch-over and LPD readings. Landing radar range data began to come in at 3min 17sec followed by altitude data 12sec after that, adjusting estimates upward by 4,800ft (1,463m) with convergence of data occurring quickly after the spacecraft cleared the top of the Apennine Mountains.

ABOVE A familiar Florida sight, lightning in the sky the day before launch with strikes close by LC-39A. *(NASA)*

Throttle down at 7min 22sec took Falcon into the approach phase and P-64 at 9min 23sec. At pitch-over the landing site ahead lay visibly similar to the simulators the crew had practised on for hours but this time the effects of the descent engine on the fine dusty particles would raise an obscuring cloud. Descending almost vertically in the landing phase from a height of about 200ft (60m), Scott and Irwin saw dust snaking along the surface seconds later and from a height of 59ft (18m) the surface was impossible to see. Final touchdown would be on instruments as it had been with Apollo 12 but with worse visible conditions. By the time of almost vertical descent Scott had already translated slightly north-east of the predicted touchdown spot to select a safer surface.

Telemetry showed that Scott made 128 separate deflections of the hand controller to move Falcon for a suitable touchdown, which occurred 1,110ft (338m) uprange and 1,341ft (409m) to the north, a straight-line distance of 1,800ft (550m) north-west of the pre-mission target. The landing phase programme under P66 began at 11min into the descent with engine shutdown at 12min 18sec when the three sensor probes brushed the surface. The right leg and forward foot pad touched the surface 1.7sec later followed by Falcon settling down into the loose regolith at 12min 21.7sec. Falcon was pitched back 11°. Scott had been efficient in landing Falcon and telemetry showed a low-level warning light came on right at the moment of touchdown, indicating that 1min 51sec of 'flying' time remained.

The general pattern of first sleep, then eat, then EVA was set down as the most effective way to manage three Moonwalks within the 67hr it was planned that Falcon should remain on the surface. But the first task for the crew on the Moon was for Scott to prepare for a unique event in the schedule, one which would not be repeated on the following two J-series landings: he was to open the upper docking hatch of the Lunar Module and, standing on the cover to the Ascent Propulsion System, visually describe the surface features, familiarising himself as well as Mission Control with the precise position of the LM relative to the features both he and Irwin had intensively studied.

It had taken a lot of hard work on Scott's part to get the Stand-up EVA (SEVA) in the flight plan. When first broached about the idea, management was reluctant to accept it but lobbying by the geologists and insistence on Scott's part swung the day. It has to be remembered that by this time in the programme there was a reluctance to take any risk whatsoever, many engineers believing that Apollo was a minimally safe endeavour and that to consistently keep doing it increased the potential for a disaster – against this background the reluctance to approve a SEVA was just one of the several 'push-backs' when additional tasks were added.

And there was sound reasoning for this. The upper hatch had to be opened and closed in a 1/6th gravity environment where it had never been operated in space other than in a weightless state. Lowering it and lifting it back to effect a tight seal, even against this reduced force on the Moon, was another potential risk. Some reasoned that it could cause an early return from the surface if a pressure integrity check revealed some leak of internal pressure after the hatch closed. But this was merely one among several cautionary concerns.

Again, when Scott argued to take a 500mm telephoto lens along for the Hasselblad camera so that he could obtain very clear and close-up shot of the hills and mountains from the SEVA it was denied at first until cleared at a very senior level – and only then by reducing the quantity of abort propellant available to compensate for the additional weight.

Dave Scott was probably the first astronaut to show this level of intense interest in the reason why they were there, placing priority on establishing the mission as a benchmark for lunar science and putting himself in the role of explorer rather than astronaut. Post-landing preparations to stay for nearly three days took slightly longer than planned, and the depressurisation for the SEVA did not take place until 4hr 00min 20sec after landing. And then Scott was in his element.

SCOTT: 'All of the mountains around here are very smooth. The top of the mountains are rounded off. There are no sharp jagged peaks or no large boulders apparent anywhere. The whole surface of the area appears to be smooth with the largest fragments in the wall of Pluton…[one of the craters marked out for identification]. There are no boulders at all on St George [a large crater about 2ml [3km] across and about 2.5ml [5km] to the south on the slope of a mountain range], Hill 305, Bennett, or as far as I can tell looking back up at [Mount] Hadley. Hadley's sort of in the shadow. It's a gently rolling terrain completely around 360 degrees, hummocky much like you saw on [Apollo] 14…I can see, as I mentioned before, Chain…and Pluton, very rounded subdued craters. It looks like the southern rim of Pluton is on the same level as our location here.'

In the science support room, geologists listened to the verbal description of the craters and features arbitrarily named for this flight; many of the places and depressions called out were too small to have been officially mapped and named. From the Capcom console, Joe Allen enthused with the crew over what they could see. It all augured very well for a significant increase in scientific activity.

SCOTT: 'I can see the Sun from the other side of the north rim of Pluton. All of it very flat, smooth and gently rolling. Inside walls of Pluton are fairly well covered with debris. Fragments up to, I'd estimate, maybe two to three metres [6.5-9.8ft]…just sort of scattered around. I look on around, and our Mount Hadley itself is in shadow, although I can't see the ridge line on the top…it too is smooth. I see no jagged

LEFT Lift-off! AS-510 launches into a clear sky on 26 July 1971, the first of three J-series missions to realise the potential with Apollo hardware and to demonstrate operational maturity. *(NASA)*

peaks of any sort. The hill I would call number 22 on your map, far distance, also looks smooth and rounded, no prominent features. I'll skip the distant field to my 6 o'clock because it's all in the shadow and looking into the Sun, of course obliterates everything.'

The Falcon's Commander was moving round in a clockwise fashion, describing things directly in front, the 12 o'clock position, round to the rear,

BELOW Crowds gather to watch this historic mission – the first to carry a car to the Moon and to provide three working days exploring the lunar surface. *(NASA)*

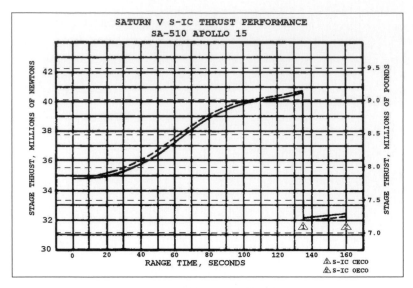

**SATURN V S-IC THRUST PERFORMANCE
SA-510 APOLLO 15**

ABOVE The thrust curve for the Saturn V showing the levels of first-stage output increasing as it moves through Earth's increasingly rarefied atmosphere. *(NASA)*

or 6 o'clock area. But all the while, personnel in the science supports room at Mission Control in Houston were poring over lunar surface EVA maps, trying to fix the LM's precise location.

SCOTT: 'As I look down at my 7 o'clock I guess I see Index Crater here in the near field. But back up on Hadley to the east of Hadley Delta again I can see smooth surface; however, I can see lineaments. I'll take a picture for you, there are some very interesting…there appear to be lineaments running – dipping – through to the north-east parallel and they appear to be maybe three per cent to four per cent of the total elevation of the mountain. Almost uniform. I can't tell whether it's structure or internal stratigraphy or what, but there are definite linear features there dipping to the north-east at about…30 degrees. As I look up to Hadley Delta itself, I can see what appears to be a sweep of linear features that curve around from the western side of Hadley Delta on down to Spur down there. And they seem to be dipping to the east at about 20 degrees. These are much thinner lineations on the mountain than I saw before, these probably are less than one per cent of the total elevation of the mountain. The craters on the side of Hadley Delta are relatively few.'

The value of this early view from the top of Falcon had great advantage in that subtle blends of lineations and stratigraphy within the surrounding hills and mountains were not visible at the slightly higher Sun elevation by the time Scott and Irwin went outside for their first EVA.

But, while the scientists were greatly enthralled by these field observations, flight controllers wanted to better understand the proximity of these features to the Lunar Module so that they could refine plans for fine-tuning the EVAs, which on these J-series flights were focused to a much greater extent on traversing the surface.

The SEVA report helped close the gap between the best resolution from the Lunar Orbiter images and the reality of rocks, boulders and obstructions through surface undulations impossible to see in those views from orbit. None of the previous orbital ground tracks on previous Apollo flights had flown over these high latitudes and there were no views from hand-held photography such as that conducted on Apollos 11, 12 and 14 which had relevance to later missions insofar as surface resolution was concerned.

SCOTT: 'Okay, coming on around to St George, which again is a very subtle old crater, but in this case I can see some lineaments running, dipping to the west at about 20 degrees parallel to the rim of the crater. The rim of the crater is very subdued and smooth. Coming around I'll just take a quick look at the near-field for you here. It's about generally the same. The crater density is, I'd say, quite higher, somewhat higher, than I expected. Sizes are mostly less than about 15 metres [49ft]. The only large crater that I can see is what I believe to be Index back here about the 8 o'clock and it has a very subtle rim, almost no shadow at the bottom of it. I think that's one of the things that was deceiving on the descent. There are very few deep, dark craters in this area… Trafficability looks pretty good. It's hummocky, I think we'll have to keep track of our position but I think we can manipulate the rover fairly well in a straight line and I can see the base of the Front…Looks like we'll be able to get around pretty good.'

The SEVA lasted 33min from depress to repress of the cabin atmosphere and the one and only area survey of an Apollo landing site from the top of the Lunar Module had been completed, without any problem sealing the hatch. As per the Flight Plan, it was time to get some sleep and make ready for a busy day on the surface. The first sleep session on the Moon began about five hours after landing and lasted around

five hours in total. EVA formally began at cabin depressurisation, 119hr 39min 10sec into the mission or 15hr 56min 41sec after landing, a little earlier than anticipated in the schedule.

The formal start was indicated by the start of depressurisation but EVA times noted in official reports throughout the US space programme have different definitions of when to indicate events; sometimes at the start of depressurisation and sometimes when the hatch was opened and, by definition of either, the duration of the event. In this instance it was marked when the pressure began to fall in Falcon. However, Irwin had a problem with the feedwater pressure in his PLSS but this did little to delay activity. Scott descended the ladder to become the seventh man on the Moon, deploying the MESA and displaying these historic steps to a global audience.

When Irwin joined Scott the two set about preparing and deploying the rover but first the TV was set up on a tripod and aligned so as to show activity around Falcon and then the ubiquitous contingency sample was bagged. In Houston it was a mid-morning TV show and not a few employees, wooed back to Moon missions by the excitement of this greatly expanded capability, watched as the two Moonwalkers set about retrieving the LRV, 95 days after it had been packed into Quad 1 on

the Descent Stage. But that took a bit longer than expected and 1hr 45min had elapsed since the EVA began before it was fully set up and configured for operations, compared with the planned 45min. There was a problem in that the front steering was not working, despite repeated cycling of the appropriate switches and circuit breakers.

The entire operation with the LRV called for Irwin to return inside the LM and transfer an EVA pallet from the surface which contained

ABOVE Mission Control, Houston. The scene in the Mission Operations Control Room (MOCR) at the Manned Spacecraft Center in Houston just 15 minutes after lift-off. *(NASA)*

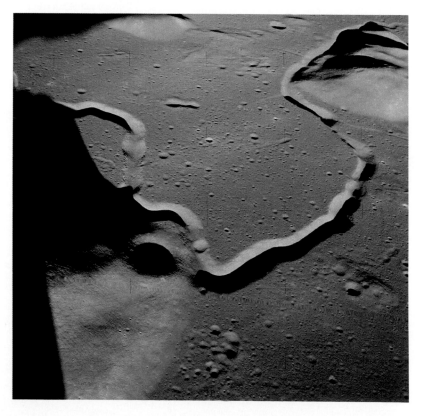

RIGHT A photograph of the landing site taken by Jim Irwin in Falcon on orbit 13 from an altitude of 7.5ml (12km). *(NASA)*

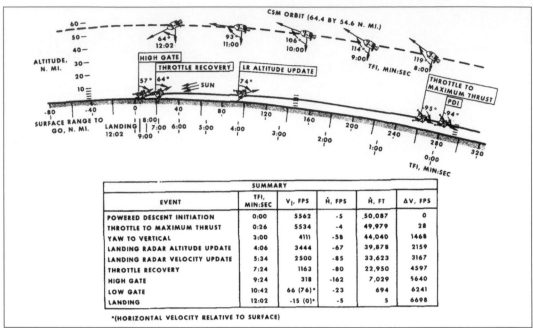

RIGHT The landing events chart showing the somewhat modified descent profile for the J-series flights from PDI to High Gate. *(NASA)*

EVENT	TFI, MIN:SEC	V_I, FPS	Ḣ, FPS	H, FT	ΔV, FPS
SUMMARY					
POWERED DESCENT INITIATION	0:00	5562	-5	50,087	0
THROTTLE TO MAXIMUM THRUST	0:26	5534	-4	49,979	28
YAW TO VERTICAL	3:00	4111	-58	44,040	1468
LANDING RADAR ALTITUDE UPDATE	4:06	3444	-67	39,878	2159
LANDING RADAR VELOCITY UPDATE	5:34	2500	-85	33,623	3167
THROTTLE RECOVERY	7:24	1163	-80	22,950	4597
HIGH GATE	9:24	318	-162	7,029	5640
LOW GATE	10:42	66 (76)*	-23	694	6241
LANDING	12:02	-15 (0)*	-5	5	6698

*(HORIZONTAL VELOCITY RELATIVE TO SURFACE)

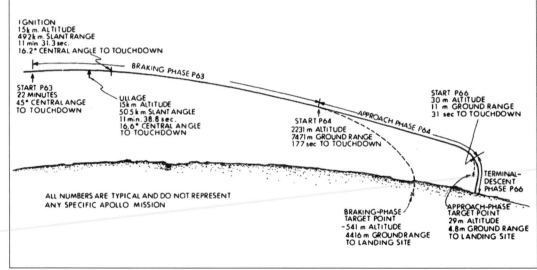

RIGHT This trace of landing events shows the automatic trajectory curve for the programmed braking phase under computer programme P63, with intervention by P64 to reject the short-range targeting and present an approach phase flight path to the point where the Commander selects P66 for manual descent to the surface. *(NASA)*

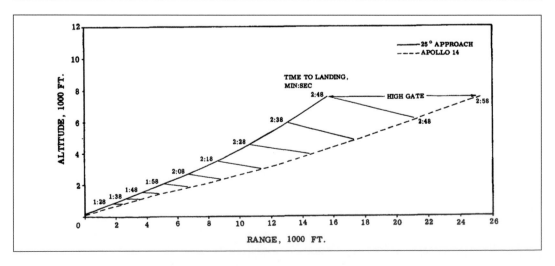

RIGHT This chart shows the approach phase from the point where it intercepts High Gate at a steeper descent angle than was the case with the Apollo 14 mission, where the descent path was over relatively flat terrain. *(NASA)*

additional food, batteries and lithium hydroxide canisters carried outside and re-configure the communications to go through the LCRU on the rover. By this time Scott had removed the TV camera and placed it on the LRV ready for the geological traverse. Unlike the H-series flights, J-series missions required the crew to conduct a geological survey before returning to the LM to deploy the ALSEP, a switch in priority due to the need to verify that the rover worked and to satisfy the requirements of the scientists, prioritising the field survey and initial sampling. Not before an EVA elapsed time of 2hr 05min did the crew set off for station 1, down to the south-west and along to the edge of Hadley Rille before moving along to station 2 near St George Crater.

During the traverse Scott discovered the peculiar characteristics of the rover, finding that it induced a bouncy ride at speeds of 5–6mph (8–10kph). After driving about 1ml (1.6km), Scott and Irwin arrived at station 1 some 2hr 30min into the EVA, where they spent 12 minutes gathering samples and taking panoramic views of the area before departing for station 2 some 1,600ft (500m) along where they arrived at 2hr 55min. More samples were collected together with a double core tube sample and some stereo-panoramic views as well as photography with the 500mm lens that Scott had pushed so hard to have included.

During these geological field activities the crew had reported excitedly on the impressive rille and viewers on Earth saw the magnificence of this feature from a slope up at station 2 where the TV camera was slewed across to show the dramatic and sinuous gorge that snaked its way north. The sharp contrast of the eastern ridge of the rille, casting deep black shadows down onto

LEFT For the first time, Apollo captures the magic of a lunar landscape replete with craters, hills and mountains, the latter missing from the relatively flat surfaces of the first three landing sites. (NASA)

ABOVE Uniquely, Dave Scott performed a Stand-Up EVA (SEVA) shortly after landing in which he opened the top hatch and conducted a visual report to Mission Control on their precise location and used that opportunity to take a sequence of photographs, his eyeline some 29ft (8.8m) above the surface. *(NASA)*

BELOW Scott described lineations in adjacent mountains, taking this shot with his 500mm telephoto lens. *(NASA)*

the narrow, winding bed, provided relief to the glistening fragments embedded in the western wall. In places the ridge appeared unnaturally sharp compared to the folding undulations of the mountains around.

Station 2 activities took 52min before they set off back to Falcon where they arrived

34min later, around 4hr 20min into the EVA but there was some noticeable drift in the LRV's navigation system. It showed 34° on a bearing back to Falcon but a visual identification of the LM showed about 15°, according to the crew. Having returned to Falcon it was time to deploy the science equipment.

Scott and Irwin next set about retrieving the ALSEP instruments from the SEQ bay in a procedure close to that followed on Apollos 12 and 14. The SEQ bay itself had a width of 54in (137cm), a height of 21in (53.3cm) and a depth of 24in (61cm). Scott drove the LRV about 300ft (91m) due west while Irwin loped along with the two sub-pallets of instruments to deploy across the surface. The greatest interest in deploying the ALSEP-4 array lay not exclusively in the intrinsic value of each item, all but one of which had been placed on the lunar surface before, but rather in the significance attached to the concurrent operation of all three arrays.

The suite of experiments contained instruments laid out on one or other of the preceding missions and included the passive seismometer of the same type deployed at Apollo 12 and 14 sites, essential to clarification of several unique phenomena observed over the preceding nine months. Prior to the emplacement of the second seismometer in the Fra Mauro formation, the recorded seismic waves which occur five days and three days before lunar perigee were incorrectly thought to occur near to the surface. Subsequent correlation of data from the two sites resulted in the discovery that these events were triggered at a depth of 50–400ml (80–650km), seeming to originate in deep pockets of magma flow. The addition of a third array helped triangulate the location of these disturbances, together with the precise location of 11 other disturbances.

The second Lunar Surface Magnetometer (LSM), the first set down by Apollo 12, would help define with greater precision the model of an asymmetric magnetic field consisting of a heterogeneous outer layer, a homogeneous inner layer and a core. For this magnetometer the only difference was the addition of a Sunshade to the top of the electronics compartment.

Also deployed at the Apollo 12 site, the Solar Wind Spectrometer (SWS) viewed the plasma sheath from a point where, at times, the Apollo

15 instrument was within the magnetospheric bow shock wave at the same time as that at the Apollo 12 site, which was external to the flow due to the different longitudinal position. The complex effects of differentiated plasma flux due to localised pockets of magnetic activity could be analysed with greater accuracy due to the simultaneous operation of the two spectrometers at the surface and with a lunar satellite, Explorer 35, in orbit since 21 July 1967. This also correlated with data from the two magnetometers at the surface.

The two SIDE experiments set down previously had already recorded strange events with spasmodic ion fields apparently flowing along lines detectable at both H-series sites. Others appeared to flare up in the vicinity of one site undetected at the other and others were believed to be propagated by the Earth's magnetosphere. The SIDE for Apollo 15 was inclined 26° to vertical to align its orientation to the same as that for those left at the Apollo 12 and 14 sites. In association with the SIDE, the CCIG was also used to correlate results with the others of its type at those earlier deployments.

Deployed to enhance triangulation of distances, the Laser Ranging Retro Reflector (LRRR) carried by Apollo 15 had 300 fused silica corner reflectors compared to 100 on those left at the Apollo 11 and 14 sites to enhance tri-axial determination of the Moon's ephemerides and the annual drift rates of continents on Earth with a tolerance of +/-2.4in (6cm).

New for the J-series missions was the Heat Flow Experiment (HFE) with a basic objective of continuously monitoring the thermal energy flowing to the surface from the Moon's interior and through the outer regolith. Two 10ft (3m) bore holes were to be drilled, 30ft (9.1m) apart, each containing two sets of ring bridge thermometers calibrated to an accuracy of 0.18°F (0.1°C) and to 0.002°F between sensors. Additional thermocouples measured the thermal flux originating from surface temperature changes while four small heaters in each hole were periodically used to monitor thermal conductivity. Each probe consisted of two 20in (51cm) string sections.

Each probe was connected to an electronics box by a 30ft (9.1m) cable. The four thermometers in each string were located

at the top of the probe and at depths of 26in (65cm), 45in (115cm) and 66in (165cm). These were spaced so as to measure the most acute change in temperature as the Moon moved from day to night, the biggest temperature gradient believed to be within the top 6ft (1.8m) of the surface and not the 3ft (91cm) at the bottom of the sensor probe.

The battery-powered Apollo Lunar Surface Drill (ALSD) used for the HFE objective was manufactured by the then Martin Marietta corporation to a specification unique in that it had to work under a wide range of environmental conditions so that it could operate against uncertain density levels in the surface. Integral to the ALSD were core bits, core stems and caps, all stowed in the SEQ bay. With a specified maximum weight of 31lb (14kg), the drill was a rotary-percussion type with operating temperatures too low to necessitate active cooling. The design incorporated tungsten-carbide bit cutters capable of operating with a small amount of dust accumulating in the bottom of the drilled hole.

With a compressive strength of 22,000lb/in² (151,690kPa) the ALSD was tested by drilling a 1.032in (2.62cm) diameter hole in

ABOVE The detail in this overview of the Hadley Apennine site clearly shows the surface features, defined by the high resolution Metric Mapping camera in the SIM bay of the Service Module. (NASA)

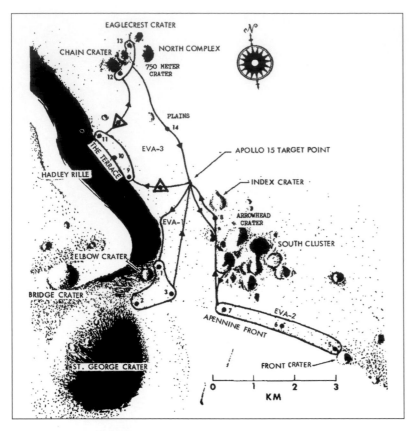

dense basalt at a maximum rate of 2.5in (6.35cm) per second. But it was accepted that the actual lunar drilling penetration rate would be degraded in proportion to the depth of the hole and the magnitude of the downward force which could be manually applied by the astronaut. Tests showed that for a hole depth of 4.9ft (1.5m) in pure basalt the penetration rate would be 1in (2.5cm) per minute.

Because the power head could reach a temperature of 250°F (121°C) a special thermal guard was attached to prevent heat transfer to the astronaut. To steady the drill, a treadle assembly provided a structural restraint during operation. Power was provided by a silver-zinc oxide battery which operated at a normal level of 23VDC at 18.75amperes for 40 minutes, the power train to the drill activated by a single-throw, heavy-duty push switch. Reduction gears ensured an output drive spindle rotation at 2,270 blows per minute and 280rpm with internal lubrication from a DuPont Krytox 143-AC oil and Krytox 240-AC grease.

The bore stems consisted of a composite of three layers of boron filament between layers of circumferentially wrapped fibreglass impregnated with epoxy. They comprised two sets of fluted boron-fibreglass tubes, one set for each heat flow probe. Each set consisted of five standard stems 21.8in (55.37cm) long and a single shorter section with a titanium adaptor. The short stem was designed so that a drill bit when coupled to the adaptor would have the same length as the other five stems, all joined by tapered locking joints. Interchangeable, the stems were each capable of attachment to the drill bit stem.

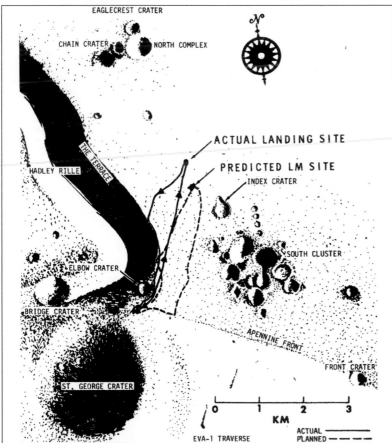

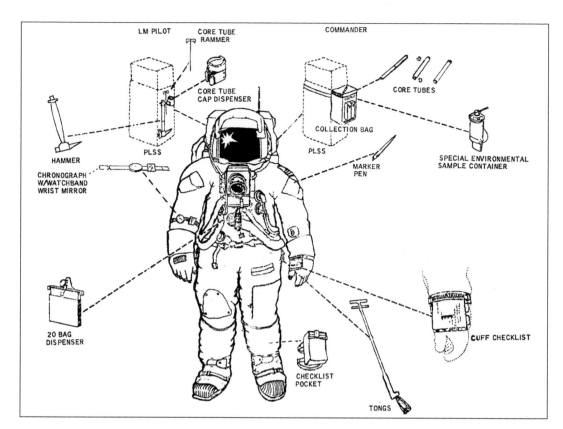

Labels within the figure:
LM PILOT
CORE TUBE RAMMER
COMMANDER
CORE TUBE CAP DISPENSER
CORE TUBES
COLLECTION BAG
HAMMER
PLSS
PLSS
MARKER PEN
SPECIAL ENVIRONMENTAL SAMPLE CONTAINER
CHRONOGRAPH W/WATCHBAND WRIST MIRROR
20 BAG DISPENSER
CHECKLIST POCKET
TONGS
CUFF CHECKLIST

When it came to the HFE task on the Moon, Scott found that the first went in to a depth of about 5.8ft (1.7m) and the probe was lowered into the hole but a second drill hole was difficult to complete and the drill got stuck. Mission Control advised him to leave that activity and begin to wrap up the EVA, electing to have Scott return to that task on the second Moonwalk. Now it was time to deploy the LRRR and the Solar Wind Collector (SWC), which would be left out until the end of the last Moonwalk, but deployment of the flag would wait for the second EVA.

But there were words of comfort for the tired Moonwalkers when Mission Control sent up a status report: 'While you are working around there you might be interested in a little conversation from down here. The SIM bay's chewing up data like it's going out of style. We're working beautifully and as far as we can determine the ALSEP is working as advertised, getting all kinds of data from it and I'll get a good accurate reading on that for you later on. And I think that your traverse goes without comment; it was beautiful and we're just trying to digest some of the data from that right now.'

The crew had driven 6.4ml (10.3km) and the first EVA was ended 30min earlier than planned due to Scott's high oxygen consumption of 1.42lb (0.64kg) from the 1.8lb (0.81kg) loaded in his PLSS, or 99.3% of the total allowable quantity for safety. This was a consumption of just 0.1lb (0.04kg) short of the redline value. At a consumption of 0.43lb (0.195kg) of his 1.78lb (0.8kg) load, Irwin was better off, having consumed 95.7% of his allowable safe quantity. But both levels were higher than expected from pre-flight calculations, energy expenditure too reaching higher levels than anticipated. Quantities of feed water and remaining suit battery power were higher than calculated too but were no problem. In explanation of this, Scott performed some of the more strenuous work, particularly during drilling, expending higher metabolic rates which required him to use the maximum setting for cooling. Irwin never went above the intermediate setting.

EVA 1 timed out at 126hr 11min 59sec, a duration of 6hr 32min 49sec, when re-pressurisation ended and the crew set about configuring Falcon for their second rest period, which began notionally at about 131hr into the mission, with Al Worden beginning his own rest period two hours later.

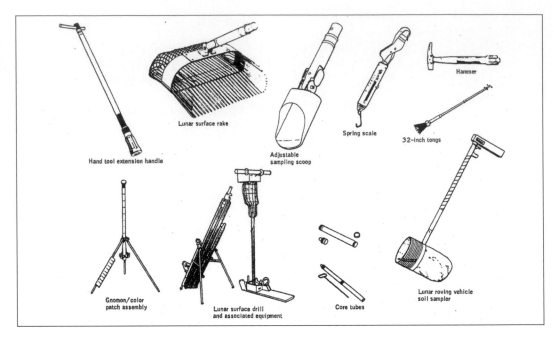

RIGHT The standard arrangement of tools and geological sampling equipment carried by the astronauts on their rock-collecting excursions. *(NASA)*

A wake-up call went out to the crew aboard Falcon on time shortly after 138hr with Mission Control advising that telemetry indicated the loss of approximately 25lb (11.3kg) of water. It was traced to a leaking bacterial filter line behind the Ascent Stage engine cover which, due to the pitch-back of the LM, had pooled in an unseen area. When that had been attended to, scooped up using bent cards redundant to further use, and towels, Mission Control advised that, based on the metabolic rates and oxygen consumption from EVA-1, this second Moonwalk would be scheduled for 6.5hr, some 30min short of the pre-flight plan.

Preparations were delayed by the need to refill the feed water tanks on Irwin's PLSS back pack. When filling those tanks after EVA-1 the PLSS had been tilted backward 30° which caused an air bubble to form and it had to be bled out before they could proceed. Cabin depressurisation began at 142hr 14min 48sec, little more than an hour later than planned, and the Moonwalkers set about preparing the LRV for its second excursion. Cycling the power switches for the front steering, suddenly it began to work and now Scott and Irwin had both front and rear available for selective use.

They now began a traverse, about 56min into the EVA that would take them far to the south, across 3ml (5km) of the Hadley plain to the 'front', a region high on the flank of Hadley Delta from where they could look back down upon Falcon's perch. Down past Index Crater they drove, turning ability enhanced with both sets of wheels steering. On past Earthlight and a group called South Cluster which they skirted to the right, avoiding little Dune Crater on their long journey south. Now they were starting up slopes that followed the gently curving lines of the Apennine front, their speed considerably reduced. Then a short rest before pressing on again. Their first sampling stop was at station 6, just beyond Spur Crater, which they reached 43min after leaving Falcon, where they gathered samples, obtained a single core tube sample, gathered a special environmental sample

BELOW Supplementary equipment carried on the crewmember's person or on the LRV. *(NASA)*

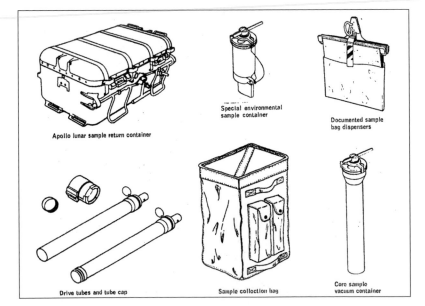

from a trench and took 500mm panoramic photographs.

Departing for station 6a, 1hr 5min later they arrived after a drive of less than three minutes, gathered samples and took some more panoramic shots. Twenty-one minutes later it was on to station 7 – Spur Crater itself – where they arrived 3hr 12min into EVA-2. There they spent 50min gathering samples, taking more pictures and obtaining a comprehensive soil sample before driving across to station 4, arriving at 4hr 15min. Sixteen minutes later they departed for Falcon. Mission Control was watching oxygen consumption rates in the backpacks and wanted them back around the LM before deciding whether to let them go to visit station 8, which had been eliminated as a priority.

Arriving at Falcon after a 22min drive they offloaded samples and configured the rover for a trip to station 8, which was close to the ALSEP site and where the heat flow experiment still required attention. The two astronauts arrived at station 8 at 5hr 7min where they gathered a comprehensive geologic sample and obtained an environmental sample from a trench.

Returning to the second heat-flow hole, Scott resumed his attempt at getting an effective depth for the sensor probe and was able to get it down to the same level as the first hole on EVA-1 but damage to the bore stem prevented further drilling.

Nevertheless, the probe in the first hole extended to a depth of 62in (152cm) from 19in (47cm) below the surface while the second reached from the surface down 43in (105cm) into the hole. After being switched on at 1947 UTC on 31 July the sensors at a depth below 32in (80cm) showed the anticipated diurnal changes and in total five of the eight sensors recorded meaningful and accurate data about surface temperatures.

Before departing, Scott was directed by Capcom Joe Allen to take a deep core sample but that too proved more difficult than expected and he was advised to leave it and return to Falcon with the aim of getting back to that on the third Moonwalk, Scott and Irwin being back at the LM at 6hr 18min where they deployed the flag, configured the LRV and climbed back into Falcon, ending a Moonwalk which had a duration of 7hr 12min 14sec. Scott had

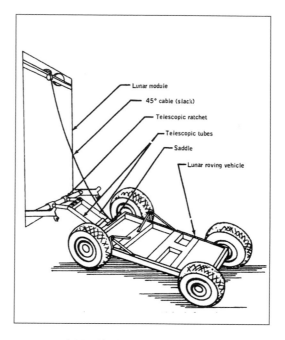

LEFT A simplified diagram of the deployment struts and booms for releasing the LRV from the Descent Stage. *(NASA)*

consumed 98.5% of available oxygen, above the redline, while Irwin consumed 80.5%, consumption of feed water being slightly higher than predicted in each PLSS and battery power also higher than calculated pre-flight.

While the crew configured their lander for a meal and the upcoming rest period, there was considerable conversation between Falcon and Mission Control regarding the work they had done on EVA-2 and about the optimum plans for the next Moonwalk. The late start and the extended duration beyond the 6hr 30min notionally assigned to the second EVA put the timeline about 1hr 15min behind but as with all

BELOW Scott and Irwin conducted a geological trip before deploying the ALSEP array of science instruments, stopping at station 1, with the rille off to the left in this picture. *(NASA)*

ABOVE Sampling at station 1, with the gnomon on the surface, a device suspended at its centre point to find local vertical with one of its legs supporting a colour chart. *(NASA)*

BELOW Dave Scott upslope at station 2 with the LRV carrying a full complement of tools. *(NASA)*

surface activity the re-configuration of the cabin always seemed to take longer than planned.

Soon the crew were 90 minutes behind time and Mission Control was working up contingency plans for Moonwalks of four, five or six hours, the pre-planned Flight Plan being for a 6hr EVA. Everything had to be worked back from the termination point for the EVA, which had to happen with sufficient time remaining for preparation for lift-off at around 171hr 37min. Scott and Irwin finally settled down for their last rest on the Moon at 153hr 15min. They had been awake for 15 hours. The plan was to let them lie for seven hours and then put in a call but the physician monitoring Irwin saw him still awake at 154hr.

The wake-up call went up to Scott and Irwin at 160hr and by the time the third Moonwalk started at 163hr 18min 14sec, the crew were 1hr 45min behind the planned time. This triggered a shortened EVA with the decision to change the traverse route and have them move in a westerly direction from Falcon straight to Hadley Rille, the impressive gorge believed to be a collapsed lava tube. The planned traverse to North Complex, some 1.5ml (2.5km) north and parallel to the terrace at the side of the rille was abandoned and there would be no excursion that far.

Leaving the LM 46min into the EVA, the first stop was near the ALSEP site just five minutes later, halting to retrieve the drill core stem samples left behind during the second EVA. Two of the sections were removed and stored on the LRV. The drill and the four remaining sections of the core stem could not be removed so they were left in situ for collection on the way back. Leaving there at 90min elapsed time they headed for station 9, where they arrived 14min later just 1ml (1.6km) away for a brief sample collection and panoramic photography session that lasted four minutes. Then it was on to station 9a, closer to the rim of the rille itself, arriving there at 2hr 1min.

They remained there for 55min, gathering an extensive variety of samples, obtaining a double core tube and performing several separate photographic tasks with included, once again, use of the 500mm camera for panoramic shots. Scott gave Capcom Joe Allen a vivid description of what he saw: 'From the top of the rille down, there's debris all the way and it looks like some outcrops directly at about 11 o'clock to the Sun line, it looks like a layer about 5% of the rille wall with a vertical face on it and within the vertical face I can see other small lineations, horizontal maybe 10% of that unit. And that unit outcrops all the rille; it's about 10% from the top and it's somewhat irregular, but it looks to be a continuous layer.'

And so it went on. For several minutes Houston listened to Dave Scott's detailed discussion of what he could see in the opposite wall and watched TV reveal a scene only moderately less precise than the commentary. But it was time to leave and at 2hr 56min they took a ride for less than two minutes to station 10 where they gathered samples and took science pictures to photograph the context of the location from where the samples were removed.

They were only there for 12min before Houston moved them on for the drive straight back to the heat flow experiment at the ALSEP site in a last attempt to get the core stems apart, leaving the rille at 3hr 10min. But as they drove across the undulating basaltic lava plain that had filled the giant basin formed nearly four billion years ago, it inspired one of the two astronauts to muse on the sight he saw:

SCOTT: 'Oh look at the mountains today, Jim, when they're all sunlit, isn't that beautiful?'

IRWIN: 'It really is.'

SCOTT: 'By golly that's just super. You know, unreal.'

IRWIN: 'Dave, I'm reminded about my favourite biblical passage from Psalms: "I'll look unto the hills from whence cometh my help." But of course we get quite a bit from Houston too.'

The two explorers were back at the ALSEP site at 3hr 27min and Scott separated one of the four connected sections of the heat flow drill but the remaining three sections were returned still assembled, which was the way they came back to Earth. The crew then returned to Falcon, arriving at 4hr 17min to park the LRV a little closer than originally planned so as to give the TV camera a better shot of lift-off. Before climbing back aboard Falcon, Scott demonstrated that under gravity all objects fall at the same rate. Using a geological hammer and a feather, he let them both go and showed that in a vacuum they reached the surface at the same time. This demonstration validated Galileo Galilei but is impossible to achieve in the atmosphere of Earth where objects such as feathers have lift which impede their fall.

On the surface they laid a small model astronaut and placed a plaque, black edged, alongside carrying the names of all 14 American and Russian astronauts who had died while in training or on a space flight, symbolising the 'fallen astronaut'.

By the time the crew were back inside Falcon

and had re-pressurised their cabin at a mission elapsed time of 168hr 08min 04sec, the EVA had lasted 4hr 49min 50sec. Abbreviated from the Flight Plan duration of six hours and the revised timeline of five hours, on EVA-3 Scott consumed 72.3% of his available PLSS oxygen, with feed water and battery power uptake also less due to the reduced time on their last Moonwalk. But it had been an extraordinary exercise in coordinated science and engineering, with emphasis very much on the former, the real reason for the J-series missions.

As he had during training sessions on Earth, Capcom Joe Allen slept when Scott and Irwin did and consistently throughout each of the three EVAs talked the Moonwalkers through the most demanding and fulfilling lunar expedition

ABOVE A view into the rille from station 2, with blocks and large boulders indicative of a collapsed lava tube. *(NASA)*

BELOW A partial panorama of the rille with the LRV and Dave Scott in attendance. *(NASA)*

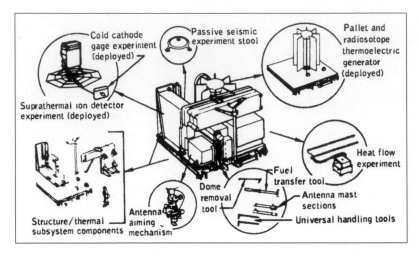

ABOVE The ALSEP subpallet No 1 with Central Station and experiments for surface deployment. *(Bendix)*

BELOW Subpallet No 2 carried the RTG, the Heat Flow Experiment and other instruments. *(Bendix)*

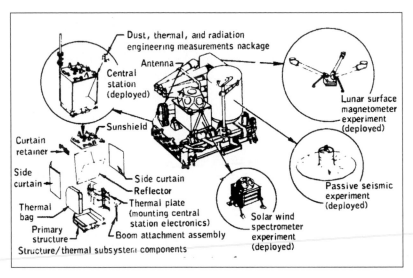

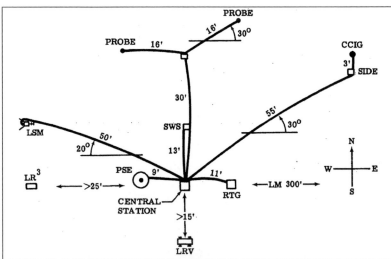

to date. But the engineers too were glued to their consoles and the amount of data analysed in the back rooms was far greater than anything on previous flights. There was still much to learn about operating space vehicles in the environment of other worlds and this extended stay on the lunar surface provided added data on the performance of the Lunar Module.

Clearly, as this was the first demonstration of the LRV, engineers were also interested in scrutinising the performance of this first motorised vehicle on the lunar surface. While there were problems with deployment and steering, as noted above, LRV-1 travelled a total distance of 17.3ml (27.9km) during an accumulated drive time of 3hr 8min on the three EVAs. Average speed was 5.7mph (9.2kph) with most of the second EVA and all of the third conducted with dual-mode steering. Maximum speed on level surfaces was 8mph (13kph). Battery consumption was less than expected, based as it had been on worst-case surface roughness and driving conditions. Estimated total consumption of 52amp-hr of the 121amp-hr available corresponds to 1.87amp-hr/km, versus the pre-flight prediction of 3.67amp-hr/km. Although there were some errors in the navigation system the closure distances were well within tolerances and the bearing back to the LM were satisfactory.

The greatest impact on learning curves was the mismatch between scheduled preparation times, sleep time and work periods, both crewmembers feeling that there were insufficient periods in the Flight Plan for all the tasks required. That, plus the shorter surface sleep times and the extended surface activity, left the crew tired and exhausted after each excursion. So much so that each man testified to being on their 'physiological reserves' at the end of each EVA.

During the Moonwalks, stretching the astronauts beyond previous experiences with earlier crews, physicians noticed irregular heartbeats in both Scott and Irwin, effects of stress they were not informed of at the time. Irwin would continue to have cardiac problems

LEFT The deployed arrangement of the experiments for the ALSEP array, laid out at the end of the first EVA. *(NASA)*

and in 1973 he had the first of three heart attacks, the second in 1986 when he collapsed on a pavement, the last in 1991 which took his life. The balance between all these physiological factors was still not in equilibrium with the wake/rest cycles of the human body and the switch from standard diurnal sleep/ wake time on Earth to delayed or advanced rest periods during space flights played hard on their physiological responses.

As a footnote, physicians came to play a considerably greater part in designing Flight Plan routines when space stations became the normal order of business for human space flight. Crews for Skylab, the programme that had taken out one of the Apollo landings to use its Saturn V for launch, were afforded a more direct approach to sleep cycles, provided with a gradual phasing of the rest periods which advanced or retreated in increments of 30min over several days so as to adjust the circadian rhythm of the human body to the time they would need to be awake for return to Earth. But such considerations were not possible with lunar flights, driven as they were by orbital alignments and the technical capabilities of the spacecraft.

During EVA-3 down on the surface, Worden conducted a Lunar Orbit Plane Change (LOPC) manoeuvre using the SPS engine in an 18.1sec burn which began at 165hr 11min 32sec, executing a ΔV of 330.9ft/sec (100.85m/sec) to align the plane of Endeavour's orbit with that of the Lunar Module coming up from the surface. Before that departure, about an hour after closing out the third Moonwalk, Falcon's cabin was depressurised a fifth time so that excess equipment could be dumped outside, reducing the mass of the Ascent Stage when it lifted off.

Ignition for that event occurred at 171hr 37min 23sec when the Ascent Stage engine ignited for a burn lasting 7min 11sec and a ΔV of 6,059ft/sec (1,847m/sec) to place the Ascent Stage in a direct-ascent orbit of 48.9ml x 10.3ml (78.7km x 14.5km). Falcon had been on its perch for a record 66hr 54min 53sec, almost three times that of Eagle on Apollo 11.

Lift-off was accompanied by a dramatic TV transmission from the rover to the strains of 'Off We Go into the Wild Blue Yonder', the signature tune for the US Air Force, played over from a taped recording in Falcon. Appropriate for an

ABOVE The Lunar Surface Magnetometer deployed at the surface. (NASA)

BELOW The electronics box for the Heat Flow Experiment is in the foreground with the RTG in the background together with the Passive Seismic Experiment. (NASA)

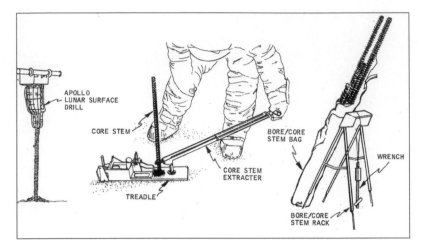

ABOVE **Essential equipment for the HFE, with treadle and core stems.** *(Bendix)*

all-Air Force crew, it was frowned upon by some in Houston as a frivolous and unnecessary adjunct when critical communications may have been necessary. The plan to have the TV camera on the rover pan upward to track the ascent was cancelled due to a suspected technical problem that may have cut the transmission had it occurred.

This fast M=1 rendezvous, first demonstrated

BELOW **The arrangement of the two holes drilled at the surface for emplacement of the thermal sensors for measuring heat flow from the Moon and to determine whether it is affected by the relative location of the Moon in its orbit of the Earth.** *(NASA)*

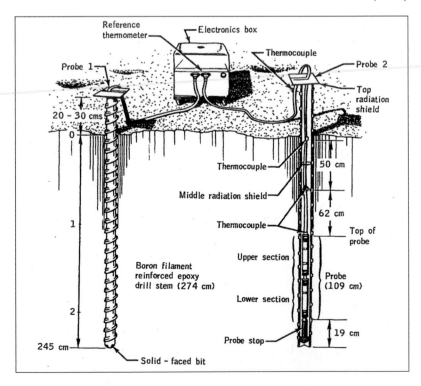

on a lunar flight during Apollo 14, required a 72.7ft/sec (22.1m/sec) second burn of the APS engine for TPI at 172hr 29min 40sec, which fired for 2.6sec on the far side of the Moon about 38min after loss-of-signal. This placed the Ascent Stage in an orbit of 74.1ml x 44.2ml (119.2km x 71.1km) and when Falcon reappeared it was 3.7ml (6km) away from Endeavour and closing at 19mph (30kph).

Docking occurred at 173hr 35min 47sec, after a separation of 72hr 56min 17sec – a period during which Al Worden had been alone and very busy with a wide range of orbital tasks. These not only involved the SIM bay and its comprehensive suite of instruments but also photography through the Command Module windows of sites of scientific interest as well as certain portions of the lunar surface in Earthshine and at the terminator – the division between reflected light from the Sun and the unlit side.

These three full 'Earth' days of migration of the terminator across the surface provided some of the most interesting photographs, registering varying levels of Sun angle as the Moon moved through fully 10% of its total rotation locked synchronously with Earth. Equipped with a much greater observing capability than any one of the previous five lunar orbit missions and remaining around the Moon far longer, six days in fact from arriving in lunar orbit to leaving, gave a glimpse of what could have been achieved with even greater and more comprehensive surveys from polar orbit, such as had been envisaged in the I-series flight which were never taken up.

Worden had already played his own part as an explorer, making extensive visual observations of special features, including fields of cinder cones which were discovered on the south-east and south-west rim of Mare Serenitatis. A landslide was defined on the north-west rim of the crater Tsiolkovsky on the far side and a zone completely free of 'rays', ejecta lines from craters and basins that lace the surface, around the crater Proclus of the west rim of Mare Crisium, thought to be a fault system. In several locations, Worden identified what he interpreted as collapsed volcanic craters seen as layers on their walls in the Maria regions.

But it was time for separating the Ascent Stage and again sending it back down to the surface to set the seismometers ringing with reverberating waves resulting from the impact. Three hours after docking Scott and Irwin were back in Endeavour, Worden's lone three-day circumnavigation of the Moon (which he testified to having enjoyed immensely) now over. When it came to undocking, a pressurisation check of the cavity that formed the docking tunnel between the closed hatches of the two spacecraft showed that pressure was decaying, indicating that either Endeavour or Falcon was leaking oxygen into that tunnel.

Delayed for a full revolution of the Moon while Mission Control had the crew open and reseat the hatch on Endeavour and verify its integrity, the crew finally got the 'go' to release the Ascent Stage, which they did at 179hr 30min 14sec, followed 19min 46sec later by a separation burn moving Endeavour away. That burn began at 179hr 50min and lasted 12.5sec as the RCS thrusters imposed a 2ft/sec (0.6m/sec) velocity change placing the CSM in an orbit of 76.2ml x 60.5ml (122.6km x 97.3km).

As planned, the RCS thrusters on the Ascent Stage fired at 181hr 04min 19sec for 86.5sec, in a 200ft/sec (61m/sec) retrograde burn to de-orbit the spent stage and send it toward the surface of the Moon, which it struck 25min 17sec later. The impact point was at 26.36ºN

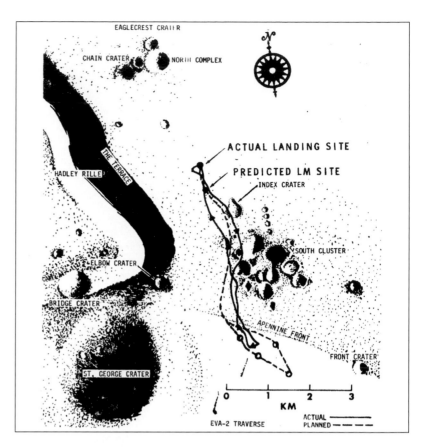

by 15ºE, 57.8ml (93km) west of the Apollo 15 site and 14.7ml (23.6km) from the targeted location. Seismometers at the landing sites for Apollos 12, 14 and 15 recorded the impact.

ABOVE The second EVA with routes modified by the slight displacement of landing spot. *(NASA)*

LEFT A pan of the boulder field at station 6A. *(NASA)*

And then it was time for Scott, Worden and Irwin to take a rest, more than three hours late compared to the Flight Plan, largely due to the two-hour delay in the succession of events beginning with separation from the Ascent Stage. It was about 183hr 15min when they finally settled down after a day lasting almost 24hr during which Scott and Irwin had conducted their last EVA, ascent from the surface, rendezvous and docking and set loose the Ascent Stage to self-destruction. Deke Slayton came on the loop to recommend that the crew take sleeping pills; the physicians had already expressed deep concern over the exertions monitored by Scott and Irwin across three days of tiring and exhausting activity. Scott declined the option.

Al Worden contacted Mission Control again at 192hr 45min and soon all three were up and preparing for a meal while listening to news from Earth – mundane but in some way relaxing, as they heard, as they had on previous days, extracts from the story of Captain Cook's ship *Endeavour*. And then it was time to prepare for a shaping-burn, a firing of the SPS engine to configure the orbit for the release of the subsatellite, still inside its container in the SIM bay. Ignition came at 221hr 20min 47sec for 3.3sec, producing a ΔV of 66.4ft/sec (20.2m/sec) putting the CSM in an orbit of 87.5ml x 62.5ml (140.8km x 100.5km). The small subsatellite was released at 222hr 39min 19sec, taking its place around the Moon while the crew made ready to return home.

One orbit later, on revolution 74, the SPS engine fired up at 223hr 48min 45sec for a TEI burn lasting 2min 41sec, adding 3,047ft/sec (928.7m/sec) and driving Endeavour out of harbour and into the ocean of space. The post-burn parameters gave the spacecraft a time at entry-interface of 294hr 57min 45sec at an entry gamma of -6.69°. A further course correction would be made but, technically, this was quite acceptable for a safe return to Earth.

After another sleep, the crew prepared for Worden to conduct his own EVA, the fifth of this mission (including the SEVA and the three Moonwalks), to go outside and retrieve the film cassette, which he did after depressurisation of the Command Module began at 241hr 57min 57sec. Less than ten minutes later the side hatch was open and Worden was on his way out the door and on his way back to first get the mapping camera cassette, handing it back to Jim Irwin standing in the open hatch who then passed it back to Dave Scott. Then it was back for the panoramic camera cassette before a final excursion to inspect a sticking mass spectrometer boom at the base of the Service Module. The EVA ended with re-pressurisation at 242hr 36min 09sec and the passive-thermal-control mode that Endeavour had been in since after TEI was resumed.

After a press conference held from cislunar space and some science experiments involving light flash observations a final sleep separated them from re-entry. After which, the last course correction of this mission came with an RCS burn that began at 291hr 56min 48sec. It lasted 24.2sec invoking a ΔV of 5.6ft/sec (1.7m/sec), producing an entry gamma of -6.49° and a calculated entry-interface time of 249hr 58min 55sec. This was just one second later than the actual time it occurred, preceded by separation of the Service Module at 294hr 44min. Splashdown occurred at 295hr 11min 53sec. At more than 12 days this was the longest Apollo mission, and the second longest US space flight to date, exceeded only by the 14-day flight of Gemini VII in December 1965.

During descent one of the three parachutes partially deflated at 6,000ft (1,828m) caused, it was concluded later, by the normal venting of residual RCS propellant during descent to prevent a hazardous and toxic environment surrounding the Command Module at splashdown. Impact rate with two parachutes deployed was 36ft/sec (11m/sec) versus 31ft/sec (9.4m/sec) for a normal landing with three full parachutes. The actual sensed impact depended upon the precise wave point when the Command Module sliced into the water. They were recovered by the USS *Okinawa*, a crew greatly relieved not to have been required to wear biological isolation garments or undergo any form of quarantine, although Dave Scott did say that he would have quite welcomed a period of rest and recuperation.

Of all the Moon missions, Apollo 15 stands out for its pioneering work on so many aspects with landmark 'firsts' identified throughout this narrative of the fourth landing on the Moon. But the enriching exuberance of the crew, the sense of exploration for its own sake and the ability to return to Earth so much more scientific data than had been possible before, singles this one mission out as the greatest of them all. More than all of that, however, was the visceral spectacle of mountains, undulating plains, a deep ravine with its snaking and sinuous trace across the lunar surface and the sheer magnificence of the valley from high up the slopes of dramatic mountains – the rim of the Mare Imbrium, the Apennine front.

ABOVE Driving back to the LM at the end of the third EVA, this view inspired Jim Irwin to quote a passage from Psalms 121: 'I'll look unto the hills from whence cometh my help.' *(NASA)*

BELOW The reduced duration of the third EVA prevented the astronauts from exploring around North Cluster and took them back to the rille instead. *(NASA)*

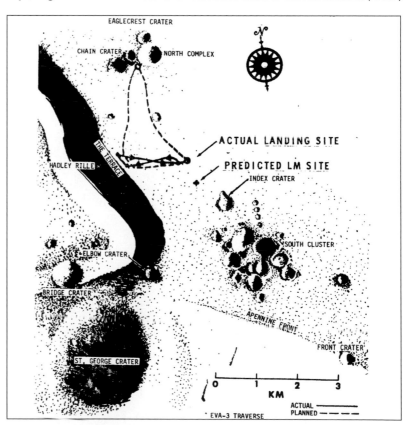

J-2 Apollo 16

Launch date: 16 April 1972
Duration: 265hr 51min 05sec

With remaining Apollo missions reduced to two per year, the penultimate J mission was scheduled for launch on 17 March 1972. By this date the future of human space flight was assured after President Nixon agreed to support development of the Space Shuttle and posed with NASA Administrator James B. Fletcher at the San Clemente, California, 'Western White House' on 3 January to publicly announce his decision. But it was a diminished configuration from the ambitious design advanced by NASA and it was widely understood that Nixon had only acquiesced against his inclination on the basis that, as put to him by a senior adviser, it was good for jobs.

For Apollo 16, the manufacturers had prepared CM-113 and LM-11 and the heavy-lift Saturn V AS-511, for which its various stages had been manufactured several years before. The crew for the mission had been publicly announced on 3 March 1971 and consisted of veteran astronaut John W. Young (GT-3, GT-X, Apollo 10) as Commander, with two neophytes, Thomas K. Mattingly II as Command Module Pilot and Charles M. Duke Jr, as Lunar Module Pilot. Young would command two Shuttle missions, as would Mattingly. They would call the Command and Service Modules Casper and the Lunar Module they would call Orion.

Back-up crewmembers included Fred Haise (Apollo 13) as CDR, Stuart A. 'Stu' Roosa as CMP and Edgar D. Mitchell as LMP (Apollo 14). All three would have formed the prime crew for Apollo 19.

The general mission characteristics were similar to those of Apollo 15 but PDI was scheduled for revolution 13 instead of 14, there would be no Stand-up EVA and there would be a rest period before lift-off. This had been ordered by the physicians, still deeply concerned over the cardiac arrhythmias experienced by Scott and Irwin during their strenuous EVAs – so much so that chief medical officer Dr Charles Berry organised a special diet for the crew which would boost potassium levels. But it was broadened to include the most stringent set of menu options yet specified in the space programme, severely restricting individual crew preferences to more comprehensively study the reaction of the human body to the expanded physiological requirements of the J-series missions.

Nevertheless, mission planners scheduled three EVAs of seven hours each instead of the 7-7-6hr Flight Plan for Apollo 15. Moreover, the ALSEP would be set out first on EVA-1 instead of a traverse first as on the previous flight. And as for the parachute failure on Apollo 15, some changes were made to the metals used on the unreefing and release mechanisms and RCS propellant would not be vented prior to splashdown.

Photographic equipment included a 70mm electric Hasselblad in the Command Module along with a 16mm Maurer motion picture camera, with three lenses of 10mm, 18mm and 75mm focal length and a 35mm Nikon F with a 55mm f/1.2 lens. Cameras in the LM included two 70mm Hasselblads with 60mm Zeiss lenses, an electric Hasselblad with 500mm lens and two 16mm Maurer motion cameras, one of which was stowed in the MESA and eventually transferred to the rover. The crew Hasselblads attached to chest mounts on their space suits.

Overall, the orbital experiments carried by Apollo 16 were identical to those on Apollo 15 but the ALSEP-D instruments were very different, including the passive seismometer, tri-axis magnetometer and Charged Particle Lunar Environmental Experiment (CPLEE) instrument

BELOW The crew for Apollo 16 with (left) its Commander, John Young making his return to the Moon and this time to land, Command Module Pilot Ken Mattingly (centre) and Lunar Module Pilot Charlie Duke. *(NASA)*

LEFT A dramatic
view of the hardware
required to get to the
Moon, with the AS-511
launch vehicle and its
associated Launch
Umbilical Tower
fixed to the Mobile
Launch Platform, with
the Mobile Service
Structure at right.
(NASA)

emplaced by Apollo 15, in addition to the same
Active Seismic Experiment (ASE) from Apollo 14
together with the thumper device and mortar
ejector package. Apollo 16 would also carry the
Heat Flow Experiment to emplace two probes
for measuring the thermal flow from the lunar
interior and to obtain accurate measurement of
surface temperatures, as had Apollo 15.

In addition, like their predecessors on Apollo
14, the crew would also carry along the Lunar
Portable Magnetometer, this time used on the
first and third Moonwalks to take measurements
of the lunar magnetic field. In addition they
would return to the Moon a sample collected
by Apollo 12 which had been 'scrubbed' of
the softer of two measured values of remanent
magnetisation, leaving a clear and stable

LEFT John Young
practises with an
engineering model of
the thumper device
which he will use
to detonate small
charges at its base
to produce seismic
waves transmitted
through the lunar
surface and picked up
by geophones. (NASA)

component. After bringing this sample back it would be analysed to see if it had re-acquired the soft component.

Unique to Apollo 16 was the Far-UV Camera/Spectroscope, a substantial device weighing 48.5lb (22kg) which was carried to the Moon inside a special zipped bag in the MESA for deployment to an area shaded by the LM. The intention was to map the concentration of hydrogen in the solar system and in interstellar and intergalactic spaces. It was the first astronomical observation conducted from the surface of the Moon. Previous attempts to employ a spectrograph in space had been marred by the Earth's corona but in observing selected areas by this instrument it was hoped to determine what many physicists believed, that hydrogen does exist between the galaxies and that it could be detected by an instrument of this kind.

Set upon a tripod and placed in the shadow, the equipment incorporated a 75mm electronographic Schmidt camera containing a potassium bromide cathode and 35mm film. The spectrograph contained lithium fluoride and calcium fluoride filters for detecting Lyman-alpha radiation in 1216 angstrom wavelength. Controls for adjusting elevation and azimuth allowed the astronaut to align the telescope with specific targets in the sky, adjusted periodically during the EVAs. At the end of EVA-3, the film cassette would be retrieved and returned to Earth, leaving the instrument on the surface.

As had been the case with all four previous Moon landings, a Solar Wind Composition (SWC) experiment would be deployed, with a plan to keep it exposed to the solar wind for around 46 hours before retrieval and return to Earth. The SWC for this mission had platinum foil strips in an attempt to reduce contamination.

Preparation for Apollo 16 was hindered by

BELOW NASA officials gather around a console prior to making a decision whether to land Apollo 16 on the Moon or to abort the landing. Seated, left to right, are Christopher C. Kraft, Director of the Manned Spacecraft Center and James A. McDivitt, Manager, Apollo Spacecraft Program Office; and standing, left to right, are Dr Rocco A. Petrone, Apollo Program Director; John K. Holcomb, Director of Apollo Operations; Sigurd A. Sjoberg, Deputy Director; Capt Chester M. Lee, Apollo Mission Director; Dale D. Myers, Associate Administrator for Manned Space Flight; and Dr George M. Low, NASA Deputy Administrator. (NASA)

THE DESCARTES LANDING SITE

The landing site for Apollo 16 had been singled out for a visit early in the programme and a final decision was made in June 1971 that this mission would go to a highland region where ancient primordial crust had been unaffected by the vast expanses of lava that flowed from the interior to fill giant impact basins, large craters and expanses of lower-lying surface features. Originally assigned to Apollo 19, cancellation of that flight shifted it up the list.

Located 37ml (60km) north of the crater Descartes, lunar scientists prioritised this site because it showed signs of volcanic activity and would allow the astronauts to visit two formations of structural material which appeared littered with deposits extruded from the mantle. Moreover, they appeared to have overlapping units that were called 'Descartes' materials and 'Cayley' materials.

The attractiveness of this site lay in the fundamental question regarding the origin of surface features and the interminable puzzle as to whether even the major craters were volcanic or impact in origin. Even world-class astronomers who had specialised in

studying the Moon for several decades were on opposing sides and the search for proven volcanic units that could be sampled and brought back for dating was uppermost in landing site selection.

The general consensus at the time of this flight was that most of the craters had been formed by impact from rocky bodies, some of tremendous size which had formed the great basins later filled with volcanic lava. But there were also indications of volcanic activity and this was what this mission was all about and why this particular site was so important.

Located between two young and very bright, rayed craters named North Crater and South Crater, the Cayley plains presented the relatively smooth landing place from which the astronauts could gather materials. The Descartes formation itself represented the hummocky, hilly surrounding features which were likely composed of igneous basaltic rock surrounded by Stone Mountain and Smoky Mountain. Some interpretations identified these as volcanic domes. The Cayley plains appeared to be relatively free of craters and to have been of Imbrian age.

some technical problems and some human problems, beginning in early January 1972 when Charlie Duke was hospitalised with bacterial pneumonia, putting back the landing to 16 April. The AS-511 launch vehicle was stacked in the Vehicle Assembly Building at the Kennedy Space

Center during October 1971, where Boeing had already delivered LRV-2 the previous month, and rolled out to LC-39A on 13 December. But on 25 January 1972 the failure of a technician to seal a pressure relief valve properly caused an RCS propellant tank in the Service Module to collapse.

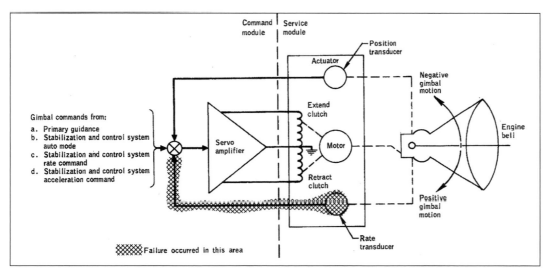

LEFT A schematic of the failure path involving the Service Propulsion System gimbal actuator and tracked to the rate transducer which had an electrical problem that caused oscillations, threatening to end the mission before the landing. *(NASA)*

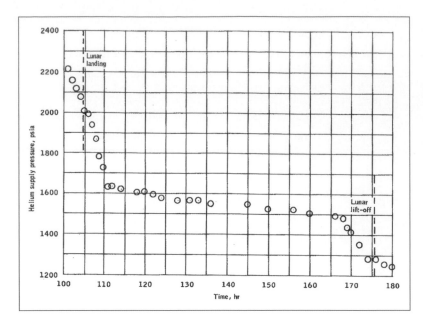

Two days later AS-511 was rolled back to the VAB for it to be repaired.

Fearing a threat to the already postponed launch date, new and fast-paced schedules saw the vehicle back at LC-39A during the evening of 9 February. Eight days after that human error again caused two pressure discs on oxidiser tanks to rupture but these could be replaced on the pad ready for a Flight Readiness Test (FRT) on 1 March. Then, two Count Down Demonstration Tests (CDDTs) were performed where all the stages involved in bringing the stack to the point of engine ignition were conducted, with the crew but no propellants on board (a 'dry' CDDT) and without the crew where the stages were fuelled up (a 'wet' CDDT). Such was the complexity of bringing every Saturn V to launch day. The final countdown to flight began six days before the scheduled launch time of 12:54:00 local time on 16 April.

Lift-off was followed by a smooth flight to orbit followed by concerns regarding the two attitude control modules on the S-IVB, a helium regulator failing on one and a helium leak detected in the second. The crew were asked to prepare to 'steer' the configuration during TLI using RCS thrusters on the Service Module in the event the helium pressurisation gas should fail completely. But that was not necessary and the remaining stack was placed on a trans-lunar trajectory at 2hr 33min 37sec followed by separation at 03hr 04min 39sec and docking with Orion just under 17 minutes later.

Separation took place at 3hr 59min 15sec,

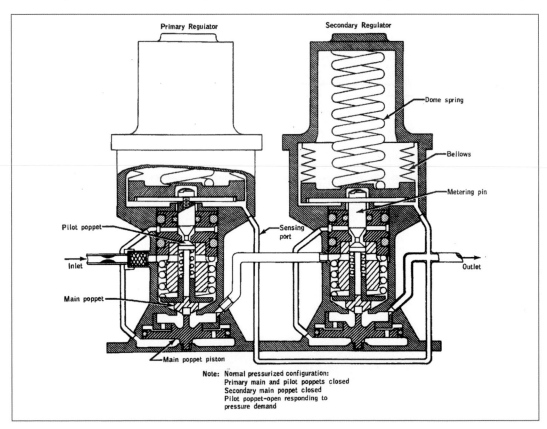

leaving an ailing S-IVB stage; the first S-IVB burn aimed for an impact point on the Moon at 2.3°S x 31.7°W but because of the loss of helium in the No 1 regulator and potential trajectory dispersions from stage venting, the second manoeuvre was not performed and lunar impact operations were terminated. Tracking was lost at 27hr 09min 07sec but the best estimate is that the stage impacted the surface at 1.8°N x 23.3°W, 96ml (155km) north of the Apollo 12 site and 683ml (1,100km) south-west of the Apollo 15 ALSEP. The magnitude of the impact allowed seismologists to profile the Moon to a depth of from 19ml (30km) to 124ml (200km).

The crew noticed a stream of light-coloured particles coming off Orion at docking and at 7hr 18min they reported a stream of particles flowing out from the vicinity of an aluminium closeout panel which covered Mylar insulation over one of the RCS systems. The panel is located below the docking target. To find out if any of the tanks were leaking, Young and Duke opened up the tunnel and entered Orion an hour later and discovered that there was no loss of consumables from any of the systems and that all quantities were as they had been on the pad. The TV in Casper was turned on at 8hr 45min to give Mission Control a view of the particles and the LM was powered down. As the spacecraft rotated out of direct sunlight the quantity of particles was dramatically reduced and the TV was turned off after 21 minutes of transmission. It would be determined later that these particles were shredded thermal paint.

On the way out to the Moon some science experiments were conducted and ultraviolet photography of the Earth took place, with a course correction at 30hr 39min 01sec for an SPS burn of 1.8sec and a ΔV of 12.5ft/sec (3.8m/sec). At spacecraft separation following LM extraction, the predicted closest approach to the Moon of 168.8ml (271.6km) imparted by the S-IVB stage at TLI had been reduced to a pericynthion of 156.7ml (252.1km) by a separation burn at 03h 59min 20sec. Now, this first mid-course correction burn settled the spacecraft into a close approach of 82.5ml (132.7km). No further course corrections were necessary.

Suddenly, at 38hr 19min 02sec the Command Module Computer (CMC) received an indication from the Inertial Measurement Unit that a gimbal lock had occurred. The computer switched the IMU to a 'course align' mode and triggered alarms. To realign the platform the CMP had to use the sextant and telescope but due to the large number of flaking thermal particles blocking his view of the stars, realignment had to use the Sun and the Earth. Believing it to be an electrical fault associated with the actuation of the thrust-vector-control (TVC) enable-relay when exiting out of programme 52, Mission Control voiced up an erasable software patch to the crew who then entered it in the CMC, instructing it to ignore indications of gimbal lock during critical manoeuvres.

The customary checkout of the LM began at 54hr 30min and was completed 41min later with all systems looking fine. A few hours later, at 59hr 19min, the docked spacecraft crossed the equigravisphere about 38,000ml (61,140km) from the Moon. The last major event prior to orbit insertion was to jettison the SIM bay door which was achieved at 69hr 59min 01sec. The LOI burn began at 74hr 28min 28sec in a burn lasting 6min 14sec for a ΔV of 2,082ft/sec (854m/sec), inserting the docked vehicles into an

BELOW Taken shortly after landing, the scene through the Lunar Module window of the Descartes lunarscape. *(NASA)*

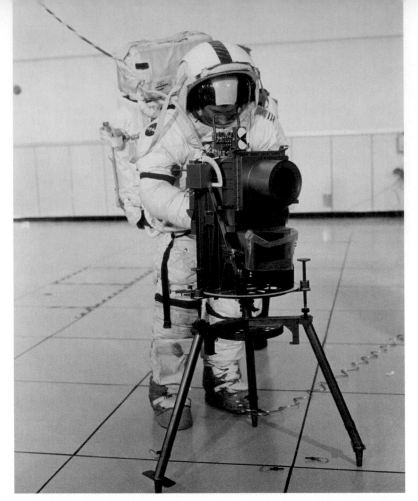

ABOVE The Far-UV Camera being 'inspected' by John Young in a publicity shot designed to draw attention to this, the first telescope operated on the lunar surface. *(NASA)*

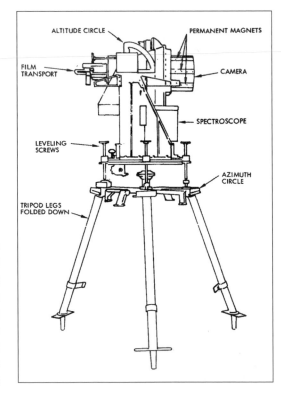

RIGHT The general design layout of the Far-UV camera designed to investigate and map the hydrogen clouds in intergalactic space and in the Earth–Moon system. *(NASA)*

orbit of 196ml x 67.3ml (315km x 108km). This was followed, as planned, by the DOI burn at 78hr 33min 45sec, a deceleration manoeuvre of 209.5ft/sec (63.8m/sec) performed with the RCS thrusters and lasting 24sec, placing the vehicles in an orbit of 67.4ml x 12.5ml (108km x 20km).

The first rest period in lunar orbit began at around 83hr elapsed time and ended nine hours later. The long rest had its dividend, the crew getting ahead of the timeline on preparing to activate Orion and start preparations for a long day – longer in fact than they imagined it would.

Young and Duke entered Orion at 93hr 34min, a little earlier than scheduled, and found a pressure rise from the RCS helium A bank pressure regulator which was unexpectedly bleeding pressurising gas into the RCS tanks. This gas was ordinarily used to fill the void between the liquid propellants and the fixed volume of the tank; as the propellants were consumed the void would increase but the leak rate was too high to leave it unattended and the crew transferred 53.8lb (24.4kg) of helium into the ascent engine tanks to reduce the pressure and relieve the RCS tanks until they could start using the thrusters during descent. After the burst disc ruptured the leak decreased from 4lb/in²/hr (27.58kPa/hr) to 1.25lb/in²/hr (8.6kPa/hr) and Orion made ready to separate from Casper, which occurred at 96hr 14min.

To get more time on the surface, the procedure for Apollo 16 had Orion start PDI on the same revolution that Casper circularised its orbit, rather than waiting a full revolution and then beginning the descent, as had happened on the previous missions where a full revolution of tracking took place before starting down to the surface. In this way, at the start of revolution 13 around on the far side of the Moon Casper would fire its SPS at 97hr 40min 17sec for circularisation into an orbit of 78.5ml x 50.6ml (126.3km x 81.4km). Orion would fire the descent engine for PDI at 98hr 35min 04sec around on the near side of the same orbit, starting its descent to the surface. It didn't happen that way.

When the two spacecraft came into view again on revolution 13 at 98hr 9min, Mission Control had difficulty getting in contact with the crew due to a problem with the steerable S-Band antenna and communications had to

go through the omnidirectional antennas. It was Orion that passed the message that Mattingly had been unable to fire the SPS engine due to a technical issue that made itself evident while he was conducting pre-ignition checklist procedures. An oscillation was noticed in the yaw axis of the secondary servo system for the SPS engine gimbal actuator but not in the primary system, which was unaffected. However, unimpeachable mission rules required that there should be a back-up in case of total failure in either system.

The oscillations were at a minor +/-1° and a rate of 2.5Hz despite the gimbal operating correctly in moving the engine in selected angles. In effect, the actuator was vibrating around a very limited deadband at medium frequency. There was suspicion that the problem was electrical because with power off the actuator the engine was repositioned without oscillation.

Until engineers came up with an answer the Moon landing was off. It would not be possible to commit Orion to a landing while the LM was still available to stand in for a failed SPS system on Casper. Of course after Orion

began its descent there would not be a back-up propulsion system for getting home anyway but the mission rules were adamant: while the LM was available it had to fill that function of a back-up. Moreover, there was only a limited amount of time in which to come to a decision.

Finite consumables were being used up by Orion and delays in landing worked on a two-hour cycle – the duration of a single lunar orbit – and flight controllers calculated that the longest they could defer the landing would be ten hours, beyond which the plane of the ground track would move steadily away from the landing site and beyond the capacity of the Lunar Module to steer out the difference and land at the Descartes site. The immediate instruction was for Orion to station-keep with Casper until a final decision was made. This was done so that, if the SPS was found to be unusable, the two vehicles would re-dock and the DPS engine on Orion would be used to push the docked vehicles out of lunar orbit for return to Earth.

As controllers huddled around the Flight Director's console manned by Gerry Griffin the lurking fear of an imminent re-dock and

ABOVE The telescopic mapping of hydrogen began when the crew deployed the Far-UV camera in the shadow of the LM to protect it from the heat of the Sun, a position from which it had to be moved, back into shadow, as the Sun got up higher. *(NASA)*

ABOVE John Young literally jumps for joy while simultaneously saluting the flag, representation of the country that built and funded Apollo and not as a national claim. *(NASA)*

BELOW The subpallet with ALSEP experiments and the thumper device for the Active Seismic Experiment. *(NASA)*

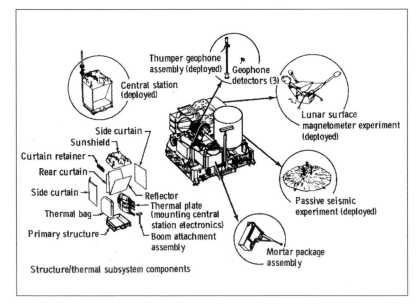

Thumper geophone assembly (deployed)
Geophone detectors (3)
Central station (deployed)
Side curtain
Sunshield
Curtain retainer
Rear curtain
Side curtain
Reflector
Thermal plate (mounting central station electronics)
Boom attachment assembly
Thermal bag
Primary structure
Lunar surface magnetometer experiment (deployed)
Passive seismic experiment (deployed)
Mortar package assembly

Structure/thermal subsystem components

immediate return to Earth using Orion's descent engine loomed large. At 99hr 15min, as the two spacecraft, now beginning to close up only a small distance apart, went around the western limb of the Moon and out of sight of Earth, engineers had telephone contact with contractors and subcontractors across the nation getting advice and consulting on possible solutions, while a barrage of communication with Ken Mattingly tried various control mode options including removing power from the hand controller to observe the response.

When the spacecraft reappeared at 100hr 4min there was a sombre tone of disappointment mixed with frustration tempered by a determination to find a solution, but the mood was given scale when the public affairs officer said that: 'The atmosphere here in the control room is reminiscent of the period just after the cryogenic oxygen tank incident on Apollo 13.' Using the rendezvous radar on Orion, the crew reported they were about 4,200ft (1,280m) apart and holding station. And from Orion, Charlie Duke expressed the view that he and John Young were thinking about not going out for the EVA before a rest as planned but flipping those two events to give them a good sleep period before the first Moonwalk.

Within about 45 minutes Mission Control had advised the crew that when coming around again to start revolution 15, the crew would be given a firm decision on whether to proceed with the landing or come home. Several procedures had been examined and there were some tentative solutions which would allow the mission to proceed but that decision had not been made when the two spacecraft went behind the Moon again at 101hr 12min. All the while, tests were being conducted at the manufacturer's plant in Downey, California, and in the simulator in

Houston, to determine whether there would be any structural problems with operating the back-up system.

The final decision rested with Manned Spacecraft Center Director Dr Christopher C. Kraft Jr, who hosted a management briefing and a technical discussion outside the Mission Operations Control Room. Engineering recommendations based on simulations at Downey gave a 'go' for using the secondary system should it be needed and confirmed that the structural integrity of the spacecraft was sound if that system had to be used. At around 101hr 45min the decision was made to go ahead with the landing and the crew were so informed at acquisition of signal some 33 minutes later. On the basis of that a second CSM separation manoeuvre was performed with the RCS thrusters at 102hr 30min to nudge Casper away from Orion and into an orbit of 68.7ml x 12.9ml (110.5km x 20.7km).

The cause of the oscillation was an open circuit in the rate feedback loop of the secondary servo system. Tests showed that if the secondary system had to be used the oscillations would damp out after ignition due to side loads on the engine bell during thrusting. Extensive examination of other vehicles in production flow did show that there was some strain on the wiring harness where it was clamped when the engine gimballed and the position of the clamp was changed so as to prevent that occurring on a future flight. There were still five Apollo spacecraft scheduled for launch including the last Moon mission, three Skylab visits and a docking flight with the Russian Soyuz in 1975.

But there was another problem, albeit minor compared to that affecting Casper. The helium gas pressurising Orion's RCS thruster clusters was now reaching a high level in those tanks because the activation procedure required pressurisation of the RCS propellant tanks before beginning the descent. But because the consecutive sequence of the CSM circularising its orbit and the LM beginning its descent had been compressed into the same orbit on this mission, Orion's crew had pressurised the tanks before knowing that the circularisation burn had not taken place. Once pressurised the thrusters had to be used in sufficient

quantity and within a relatively short space of time – as they would be for a descent to the surface – to partially consume the propellant making space for more helium.

The problem was compounded by a leaking regulator in System A, one of two redundant loops ensuring that at least one RCS loop was operable, first noticed at about 95hr GET while the crew were activating Orion. The leaking regulator would be a constant problem throughout the mission although leak rates went down to near zero at times. The most likely cause of the leak was deemed to be micro-particles, many being found in two regulator assemblies taken apart at the manufacturer's plant revealing such contaminants to be present. It merely required a single speck, 100 microns in size, to effectively block a tight seal and cause a leak.

The plan now was to complete the front-side pass and conduct the circularisation burn around the far side at the start of revolution 16 and have Orion start the descent to the surface as it reached perilune on that pass. Mattingly successfully fired the SPS engine for the circularisation burn at 103hr 21min 43sec. Almost exactly 30 minutes later the two spacecraft swept into view around the front side of the Moon reporting a successful burn, lasting 4.6 sec for a ΔV of 81ft/sec (24.7m/sec) shifting Apollo to a near-circular orbit of 78.2ml x 61.1ml (125.8km x 98.3km).

Aboard Orion, Young yawed the LM around to optimise the orientation of the omni-antenna

BELOW The subpallet No 2 with RTG and the Heat Flow Experiment, neutralised on this mission when Young caught his boot in the cable and ripped it from its connector. *(NASA)*

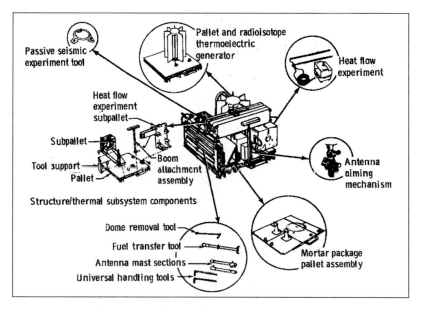

for Mission Control to upload information to the computer; the S-Band antenna was still not steering. Several thruster firings had been made to 'burn' off some of the RCS propellant and make room in the tanks' void for the leaking helium. At the rate it was leaking there would be no problem with Orion sitting on the surface not using its thrusters as the leak rate was insufficient to over-pressurise the tanks again. At PDI Orion was at an altitude of 66,500ft (20,270m) and the target was out of plane of the ground track by 4.1ml (6.6km), which Orion would steer out albeit at the cost of propellant which would slightly reduce the amount available for hovering above the surface.

The descent engine ignited at 104hr 17min 25sec for PDI which followed the by now familiar sequence of events. At two minutes the crew entered an update of 800ft downrange into the onboard computer, with altitude and velocity information coming in from the landing radar at about 50,000ft (15,240m). At 20,000ft (6,096m) Young was able to make out Stone Mountain as well as North Ray and South Ray craters and at 14,000ft (4,267m) the entire landing site was visible, indicating a very precise approach phase.

Pitch-over for High Gate was marked at 7,200ft (2,194m) and further observations with the LPD showed that the auto-land programme was taking Orion 1,970ft (600m) north and 1,312ft (400m) west of the centre of the landing ellipse. To compensate, Young began to

BELOW A diagram of the ALSEP arrangement with the geophone string and mortar line of fire shown. (NASA)

re-designate the landing spot through manual manoeuvres but there was no requirement to touch down precisely at that spot and at 450ft (137m) Young could see the shadow of the LM moving across the surface. Small evidence of dust began to appear at a height of 80ft (24m) and Young put Orion in a hover at 20ft (6m), slowly descending to the surface at 1.5ft/sec (0.45m/sec), or 1mph (1.6kph). Young killed the descent engine about one second after the contact light came on, the LM falling gently under 1/6th gravity to rest on the surface, in an almost perfectly vertical state, 12min 10sec after PDI.

Acknowledged by a cry of 'Wow! Wild man, look at that…Old Orion has finally hit

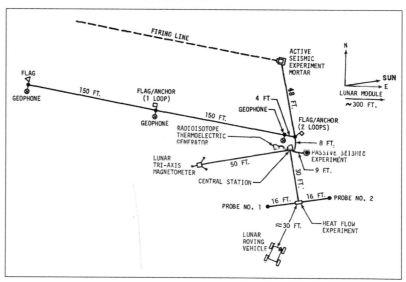

FIRING LINE

ACTIVE SEISMIC EXPERIMENT MORTAR

N

SUN
E

LUNAR MODULE
~300 FT.

FLAG
GEOPHONE

150 FT.

FLAG/ANCHOR
(1 LOOP)

GEOPHONE

150 FT.

4 FT.
GEOPHONE

48 FT.

FLAG/ANCHOR
(2 LOOPS)

8 FT.

PASSIVE SEISMIC EXPERIMENT

RADIOISOTOPE THERMOELECTRIC GENERATOR

LUNAR TRI-AXIS MAGNETOMETER

50 FT.

CENTRAL STATION

30 FT.

9 FT.

16 FT.   16 FT.   PROBE NO. 2

PROBE NO. 1

≈30 FT.

HEAT FLOW EXPERIMENT

LUNAR ROVING VEHICLE

The late landing time would force an increase in consumption due to the added heating effects from a higher Sun angle. Already, nearly 8lb (3.6kg) had been consumed in orbit.

The oxygen supply was the most promising of all. Under a nominal mission, nearly 50% would be available as reserve from a fully tanked supply of 98.2lb (44.5kg). But even accounting for all the consumables margins, over a scheduled 73hr stay on the surface the orbit of the CSM would have precessed too far west for the Ascent Stage to steer to a rendezvous. Although it would be some time before Mission Control announced it, in the Mission Operations Control Room it had already been decided that there would be a reduction in the stay time of about six hours.

After a meal, their sleep began at 108hr and ended at 115hr 30min with a meal and preparations for EVA-1. The Orion crew had conducted a detailed description of the general area around the LM which helped flight controllers and route planners to fine-tune the first Moonwalk, which began officially at 118hr 53min 38sec. Young and Duke got LRV-2 offloaded and deployed by an EVA elapsed time of 39min. Then the Far-UV camera was set up and the TV activated 1hr 12min into the Moonwalk after it had been attached to the LCRU on the rover; without that there would have been no TV transmission from their surface activity.

The first duty was to deploy the ALSEP instruments, the two sub-pallets offloaded at 1hr 29min. It took 2hr 34min to get all the equipment laid out. The heat flow experiment had gone well, a hole being drilled to the depth of the first bore stem in 51sec, an average rate of 5.8in/sec (2.3cm/sec), followed by the second stem section added and drilled down in a further 39sec. When added to those, the final section was drilled down in 54sec to a depth of 98.4in (250cm).

But the difficulty of moving around the suite of instruments with their respective data cables, each with its own coiled 'memory', connected to the Central Station became all too evident when John Young inadvertently caught his foot in the flat cable for the heat flow equipment, tearing it loose from its attachment point at 3hr 28min into the EVA.

it, Houston. FANTASTIC!', the landing had occurred 5hr 42min later than planned, a period during which Orion had been eating up its finite life-support and systems-support consumables. It had also been decided to swap EVA-1 for their rest period which, delays homologated, would put that first excursion back 17hr on the original timeline. After receiving a 'stay' for T-1 and T-2, the first job was to power down the LM to conserve power and preserve the planned duration of the surface activity before lift-off. With consumables having been used in lunar orbit, which would ordinarily have been expended on the lunar surface, doubts crept in on the sufficiency of oxygen, water and battery power to last for the planned stay time.

With 1,970amp/hr of electrical power available from the five batteries in the Descent Stage the nominal usage profile allowed a 324amp/hr margin at the conclusion of the planned stay. However, a little more than 150amp/hr of this had already been used during the wait but prospects looked reasonably good for power supplies. Water consumption had been budgeted to allow a usable 22lb (10kg) remaining from a 370lb (168kg) supply on a nominal stay and this would be marginal.

RIGHT Attention at the LRV as the Moonwalkers press on with geologic field activity. *(NASA)*

Nothing could be done, a repair impossible on the surface of the Moon, and the planned second drill operation was abandoned, the HFE now a totally failed experiment. But a successful deep core was obtained to a depth of 8.5ft (2.6m). Then it was time for Young to go about the Active Seismic Experiment with its geophone array and the 'thumper' device which operated much as that on Apollo 14 had done with good seismic data being recorded at the geophones.

At 4hr 5min Young and Duke left for their first geological field trip and a visit to station 1 across the Cayley plains to Flag and Plum craters, arriving 26min later. They found the terrain hilly and hummocky with blocks three feet (1m) in size ranging in density from 10% to 50% of the surface. Leaving 51min later (EVA-1 time of 5hr 17min) they stopped at Spock and Buster, two craters where they did more sampling and took a measurement with the lunar portable magnetometer, departing after 27min.

Station 3 was back at the ALSEP site where they arrived at 6hr 1min to perform a test of the LRV, a veritable 'Grand Prix' in which the rover's performance was assessed against some pre-planned turns, accelerations, braking stops and a few surprise manoeuvres with Charlie Duke conducting a running commentary.

DUKE: 'Man, you are really bouncing around… He's got about two wheels on the ground. It's a big rooster tail out of all four wheels, and as he turns he skids. The back end breaks loose just like on snow. Come on back John…Man, I'll tell you, Indy's never seen a driver like this. Hey when he hits the craters it starts bouncing, it's when he gets his rooster tail. He makes sharp turns. Hey, that was a good stop, those wheels just locked!'

After nine minutes at the ALSEP, a couple of which were putting the LRV through its paces, they drove back to the LM, deployed the Solar Wind Composition experiment, got back inside and re-pressurised Orion at an elapsed time

of 7hr 11min 02sec, a mission time of 126hr 04min 40sec. Apart from the abandoned Heat Flow Experiment, the first EVA had gone well, with consumables within expected limits and few surprises, one of which was a failure of the rear steering assembly on the LRV until later in the mission when it started working in time for the Grand Prix! In total on this EVA, LRV-2 had been driven 2.6ml (4.2km). The sleep period

BELOW The Descartes site with local geological features and names identifying structures and craters which will be visited on this mission. *(NASA)*

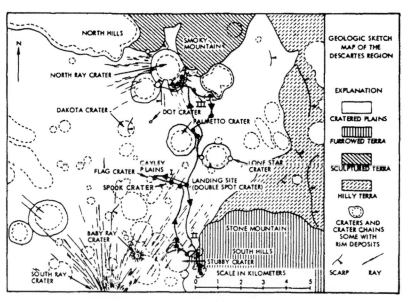

BELOW Almost lost in the flat featureless surface, the Heat Flow Experiment probe, put in place but useless after Young tripped on the data cable. *(NASA)*

for Young and Duke began at 131hr 30min and ended a little over seven hours later.

EVA-2 began at 142hr 39min 35sec when Orion's cabin was depressurised and the LRV was prepared for a geological traverse to the flanks of Stone Mountain, leaving the vicinity of the LM at 52min into the EVA, driving south for station 4 on a heading of 164°, down past an incline called Survey Ridge with large cobbles up to 12in (30cm) in size, then through blocks and boulder fields before rough, undulating terrain as they gradually climbed up the slopes of Stone after a drive of 36min.

At station 4 they shot 500mm telephoto views, raked the soil for pebbles and small rocks, jabbed the surface with a penetrometer to measure its hardness, sampled a glass-splattered bead, took core samples with the hammer and scooped up regolith. They spent just under an hour there before moving back north to station 5 more than 1,640ft (500m) away and spent a further 48min obtaining a rake sample from a crater wall, obtained a further reading from the portable magnetometer, took panoramic photographs and picked up a large crystalline rock. Then it was an eight-minute drive north-west to station 6 where they arrived at 3hr 27min to gather samples and get another set of panoramic photographs.

Departing for station 8 (7 had been deleted to give more time at station 10) after 23min at station 6, they drove for 11 minutes to arrive on the flank of Wreck Crater where they would spend more than an hour, collecting fragments of a shattered rock, scooping up regolith, completing a documented sample (so called because it required an extensive sequence of photographs to fit the retrieved material within its local context) and conducted a quick analysis of the rear steering which had been a problem on EVA-1, discovering a mismatch of switch positions to have been the problem.

Young and Duke left station 8 at 5hr 9min, only a little further north and a five-minute drive to station 9, where they were to spend 36min, still some 1.6ml (2.6km) from Orion. After overturning a boulder for sub-samples, putting glass crystals in a collection bag and taking more pictures, they packed for the traverse back to station 10, a position midway between the ALSEP and the LM where the crew was given a 20min extension on their EVA. Penetrometer readings were taken there, where they had arrived at 6hr 15min, as well as a double core tube sample and additional panoramic photography. At 6hr 40min Duke examined the HFE data line pulled from its mounting on the previous EVA; a troubleshooting effort on Earth had indicated there was a possibility of repair but to do so could have threatened the integrity of the other experiments. Back in Orion, EVA-2 had lasted 7hr 23min 09sec, a mission elapsed time of 150hr 02min 44sec.

Officially, their sleep period began at 155hr but conversation continued on for some time with Ken Mattingly in Casper over procedures required to secure a good SPS burn in view of the problems with the secondary control system before Orion was cleared to land. That, as well as some different procedures for the

rendezvous and docking, now significantly affected the nominal timeline. The call up to the crew via Capcom Donald Peterson at 162hr 50min started the last day on the Moon which included a shortened EVA from a pre-planned seven hours to five hours. This was due to the late landing and the need to shorten the stay time to keep the orbital plane within accessible reach by the Ascent Stage.

EVA-3 began at 165hr 31min 28sec, some 30min earlier than scheduled and with an abbreviated excursion to a modified set of station stops. Departing the vicinity of Orion 38min into the EVA, they drove north to station 11, right on the rim of North Ray Crater some 3.3ml (5km) from the Lunar Module, arriving 35min later. Along the way they crossed ridge lines that seemed so subdued when seen from a distance, distinct barriers close up, and across blocky rock fields strewn over the regolith. North Ray was approximately 0.6ml (1km) across two massive boulders, one about 65ft (20m) long and 16.4ft (15m) high and the other, 49ft (15m) high, appeared to merely sit on the rim.

North Ray was spectacular and the two explorers set about their work with rake samples, 500mm telephoto shots of Smoky Mountain further north and selective rock sampling before loping back around to the

'house sized' rock, impressive by its sheer size
and location right on the very rim of the crater.
Young and Duke left station 11 after 1hr 25min
and drove to station 13, a short eight-minute
drive where they arrived at 2hr 46min, 2.4ml
(3.8km) from Orion. Here they spent 29min for
more sampling and portable magnetometer
readings, moving on to station 10' (10-prime), a
newly inserted stop making a triangle with the
old station 10 and the ALSEP, arriving there at
3hr 43min for a 16min stop, getting a double
core sample and photographs. Then it was time
for a final 'Grand Prix', setting up a lunar speed
record of 10.6mph (17kph).

Back at the LM by 4hr 30min, Young and

Duke retrieved the Solar Wind Composition
experiment, rolling it up for return to Earth,
performed two portable magnetometer
readings, gathered local samples and began
the closeout activity before retrieving the film
cassette from the Far-UV camera, which was
left in the shadow of Orion and had to be
moved twice to keep it out of the encroaching
Sun. The EVA ended and the LM was
re-pressurised at 5hr 40min 03sec – a mission
elapsed time of 171hr 11min 31sec.

Before a sleep period could begin, Young
and Duke had to remove their gloves and
helmets, which Duke found difficult after a
leaking orange juice drink bag had emptied

its contents down his neck, binding the helmet connecting ring like a glue. Then, at about 172hr 40min, after hooking up the suits to the environmental control system, they depressurised the LM and discarded unwanted items out the hatch.

Procedures for rendezvous and docking were broadly similar to those of the previous two missions. A plane change was performed by Mattingly in Casper at 169hr 05min 52sec, an SPS burn of 7sec for a ΔV of 124ft/sec (37.8m/sec) placing the CSM in an orbit of 75.5ml x 63.3ml (121.5km x 101.8km). Orion left the lunar surface at 175hr 31min 48sec after a total stay-time of 71hr 02min 13sec for an ascent burn of 7min 27.7sec. A tweak burn was performed 10min 30sec later to adjust for the Ascent Stage being 6.5ml (10.4km) further downrange than planned, followed by the second firing of the ascent engine for 2.5sec at 176hr 26min 05sec, starting the terminal phase of rendezvous. After a couple of correction burns to effect rendezvous, docking was achieved at 177hr 41min 18sec.

Casper and Orion had been separated for 81hr 27min 18sec, compared with a pre-mission plan of 77hr 56min, reduced from what it might have been by shortening the third EVA and reducing the amount of time the LM remained on the surface. The sleep period for Young and Duke began at around 181hr 45min and ended with a wake up call at 189hr 30min to start a busy 'coming home' day.

After separating the Ascent Stage at 195hr 00min 12sec, Casper conducted a separation manoeuvre with the RCS thrusters 3min 01sec later. An erratic attitude-hold caused by the crew leaving some switches in the wrong position prevented the planned de-orbit burn and the stage remained in lunar orbit with a lifetime of about one year. Few tasks remained to be done and the second and last subsatellite deployed on a J-series mission was released into lunar orbit at 196hr 02min 09sec. It would eventually impact the lunar surface on 29 May 1972 after 425 revolutions of the Moon.

Trans-Earth Injection began at 200hr 21min 33sec, the SPS engine firing for 2min 42.3sec for a ΔV of 3,371ft/sec (1,027.5m/sec) producing a predicted entry interface of 265hr 36min 52sec but an entry gamma of -7.44°.

Two very minor course corrections were applied on the way back, the first at 214hr 35min 03sec for a ΔV of 3.4ft/sec (1m/sec) and one at 262hr 27min 31sec for a ΔV of 1.4ft/sec (0.42m/sec), each with the RCS thrusters and both producing an entry gamma of -6.5°.

Between the two course correction burns, Mattingly performed a 1hr 23min 42sec deep-space EVA starting at 218hr 39min 46sec using similar procedures to that conducted by Worden on Apollo 15, retrieving cassettes from the SIM bay. Splashdown occurred at an elapsed time of 265hr 51min 05sec, some 24.5hr shorter than the pre-flight plan, the crew recovered by the USS *Ticonderoga*.

ABOVE Young sampling a house-size rock at station 11. *(NASA)*

LEFT Panels of insulation shaken loose and detached from their fittings on the back of the Lunar Module Ascent Stage brought modifications for the last Lunar Module to fly to the Moon. *(NASA)*

J-3 Apollo 17

Launch date: 7 December 1972
Duration: 301hr 51min 59sec

The Apollo 17 mission may very nearly have been assigned to a lunar polar orbit objective, devoid of a Lunar Module and committed to a total global survey of the Moon rather than a landing. Successor to George Mueller, who had retired from NASA on 10 November 1969, NASA manned space flight boss Dale D. Myers was supportive of a very different job for the last Apollo mission to the Moon. Reflecting on the I-series mission in Maynard's alphabet-steps to lunar exploration, he instructed the Manned Spacecraft Center to investigate the possibility of sending CSM-114 to lunar polar orbit.

The possibility of using Apollo hardware for this kind of mission had never really gone away. Swept out of the trampling path of the media frenzy surrounding Moon flights at the Manned Spacecraft Center, it was occasionally discussed and debated in bars and diners in and around Houston whenever engineers and scientists gathered 'out of house' to talk about the future and muse over possibilities. It had much merit to it.

The outstanding success of the two J-series flights to date had vindicated the modification of the Service Module to support high resolution photography of the lunar surface within the selenocentric latitudes defined by the orbital ground tracks of the Apollo 15 and 16 missions. At maximum, these got to +/-26° of latitude and, consequentially, left vast swathes of the Moon unexplored by the battery of scientific instruments carried aboard Apollo.

Clearly, getting Apollo into a polar orbit required considerably more energy than inclinations closer to the equator and the duration of the Apollo spacecraft was at most 14 days – one complete revolution of the Moon on its polar axis takes twice as long – so any survey conducted within the capabilities of the Apollo spacecraft would have to be assigned to only one hemisphere under sunlight. Given the five days to get there and back, Myers focused on a lunar orbit duration of nine days and asked MSC to examine options based on that profile.

Energy requirements for a nominal CSM/

ABOVE The crew for Apollo 17, the final manned lunar landing of the Apollo era, with Commander Gene Cernan (right), Lunar Module Pilot Jack Schmitt (centre) and Ron Evans. *(NASA)*

RIGHT Joe Engle, who would have explored the Moon with Gene Cernan but was replaced by Schmitt, the first geologist to fly an Apollo Moon mission, had great experience with the hypersonic X-15. *(NASA)*

LM mass of 89,000lb (40,370kg) were almost 2,900ft/sec (883.9m/sec) to achieve equatorial orbit; a CSM-only mass of 51,000lb (23,134kg) to polar orbit would require a ΔV of 3,200ft/sec (975m/sec). MSC put out proposals for a candidate science payload based on either the Nimbus 4 satellite or the Docking Module which was being developed for the Apollo-Soyuz Test Project (ASTP). Developed as a second-generation weather satellite, a replica of Nimbus 4 could be left in lunar orbit when the CSM came home, effectively providing high-resolution imaging for the remainder of the lunar 'day'. The Docking Module from the ASTP programme was designed to provide an environmental airlock between the different atmospheres and pressures of the Apollo and Soyuz spacecraft which were to fly a rendezvous and docking mission in 1975.

In the end, it was decided to satisfy the broader objectives of a combined lunar science and orbital survey flight sent to a geological site where the full range of Apollo hardware could be exploited. Fundamental also to this decision was the shift in emphasis within the space programme itself away from crewed missions to robotic spacecraft which could operate far longer, without life-support resources, without tiring or the need for pause and rest and at a very much cheaper and more sustainable development and operating cost. All the objectives of a lunar polar-orbiting mission could be met with unmanned vehicles.

The crew expected to fly Apollo 17 had been the back-up crew on Apollo 14: Eugene A. Cernan as Commander, Ronald E. Evans as Command Module Pilot and Joseph H. 'Joe' Engle as Lunar Module Pilot. As had been the case in most assignments, they were on rotation to get seats on the final Moon mission. But at the formal announcement of crew selection on 13 August 1971, Engle had been replaced by Harrison H. Schmitt, one of the first six NASA science-astronauts selected in June 1965, a group specifically chosen as primarily scientists rather than career pilots.

There had been tremendous pressure from the distinguished scientific bodies in the United States for NASA to start appointing missions to those science-astronauts. But the underpinning culture within the agency was for engineering

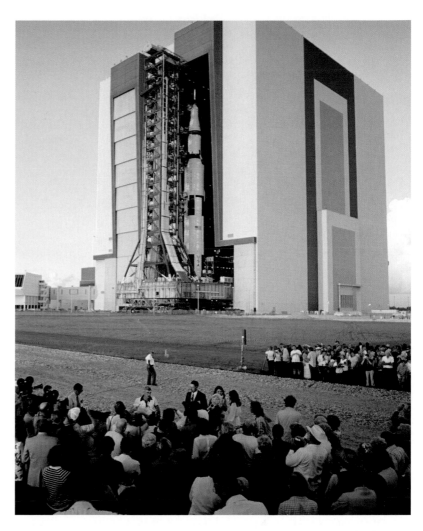

ABOVE Large crowds gather to watch the last three-stage Saturn V leave the Vehicle Assembly Building for LC-39A, Cernan (at bottom wearing blue coloured T-shirt) on hand to liaise with the press. *(NASA)*

and for qualified pilots to form the core of Apollo crewmembers, especially since it was generally considered that these flights were hazardous and required every bit of natural instinct with flying machines and complex systems management.

Joe Engle had been an X-15 pilot and had twice exceeded an altitude of 50ml (80.45km), which the US Air Force considers the demarcation line between Earth and space; the international regularity authority on such matters deems 62ml (100km) as that point. But denied a walk and a drive on the Moon, Engle would become one of the pilots for the Shuttle approach and landing tests in 1977, commander of the second orbital Shuttle flight

in 1981 and commander of the 20th Shuttle mission in 1985. But resistance to having a scientist-astronaut replace a man of Engle's impeccable flying career was boosted as well by the paucity of candidates.

So it was that Jack Schmitt got to walk on the Moon, having been pulled up to the LMP seat from his original assignment to fly on Apollo 18 at a time when other changes were taking place. In April 1972 the back-up Apollo 17 crew of Scott, Worden and Irwin was replaced by John Young, 'Stu' Roosa and Charlie Duke,

when a legal matter set aside the Apollo 15 crew for alleged malpractice when they sought financial gain from the selling of postal covers carried to the Moon.

The pace of preparation for the mission was much slower than with preceding flights and there was an overwhelming sense of finality, a door closing tight on an era of expanding human space capability defined by voyages by astronauts to another world in space. The single work-shift preparation of the hardware was a far cry from the round-the-clock preparations that kept lights burning all night, workers engaged with a very different tempo to the one that prevailed for this last Moon landing. In the weeks before the launch, many present and former workers would walk past the giant rocket, stopping to take a photograph of this last three-stage Saturn V, hardened professionals caught with moist eyes and sad looks on their faces.

It had by no means been inevitable that the Nixon administration, persuaded to allow, and to fund, the last three Moon missions, would not peremptorily pull the plug on this last flight. At a dinner at the National Space Club in Washington DC in November 1971, when asked about the cancellation of Apollo 17, NASA Administrator Jim Fletcher – who had already done so much to get approval for the Shuttle – remarked that the fate of that flight 'is beyond my control at the present time'. In fact, early on, the White House had asked NASA not to fly the mission in the November window because the President feared that a catastrophic disaster would undermine his reputation at his election for a second term. But although re-scheduled to December, it was not cancelled; it would fly and it would visit one of the most exciting places visited to date.

But there had been a concerted effort by Jack Schmitt to get Apollo 17 assigned to a landing on the far side of the Moon, where geologists knew there was a very different mix of surface features that could help explain why the two hemispheres are so different from each other. It would not be possible to set down on the far side without a separate satellite to relay communications to Mission Control and the entire architecture of the flight would have been

a radical departure from practised procedures that underpinned five safe landings thus far.

It had been an American spacecraft, Ranger 4, that became the first man-made object to impact the far side, that occurring on 26 April 1964, but it had crashed to destruction. The technical complexities of a landing on the far side of the Moon were immense at this stage of the Apollo programme and the idea was quickly dismissed – but Schmitt certainly raised the matter whenever he could. So it would be left to the Chinese to put the first robot, and a small roving vehicle, on the far side of the Moon in January 2019, supported by a communications relay satellite.

New science on the Moon

The landing site for Apollo 17 was officially announced on 16 February 1972, decided by the orbital photography of Al Worden on Apollo 15 and by his verbal verification that he had seen what he believed to be cinder cones in an area around the eastern 'shore' of the Mare Serenitatis. That focused attention to the Taurus-Littrow site, which it was agreed would be the place to target this last J-series mission. It had the added advantage of several unique geologic units within accessible driving range of the rover.

The team who began reviewing landing site selection was led by Noel Hinners, with geologists and Jack Schmitt. That review of options had begun in November 1971 and candidate sites had to take account of the relative position of the Lunar Module when it began its descent.

With a requirement for 12min of tracking as the Lunar Module appeared around from the far side, the furthest east a landing site could be was 34° and only under circumstances of exceptional scientific merit should a site be further east than 43°. But the preferred landing site was at 30°E by 20°N, just within the comfort zone for engineering the trajectory of a safe landing with bale-out zones where there would be sufficient tracking and guidance updates prior to PDI.

But there was one last stretch of the rules. In looking at places to put a Lunar Module down, the Mission Planning and Analysis Division (MPAD) had always stipulated a landing ellipse of 1.86ml (3km) long by 1.24ml (2km) wide which would have to fit within obstacles known to exist at the surface. The flat region at the Taurus-Littrow site did not have such an area and dispensation was granted for this mission with a circular area of 0.62ml (1km) permitted.

The orbital experiments installed in the SIM bay for Apollo 17 have been described earlier but the ALSEP-E surface experiments differed substantially, with only the Heat Flow Experiment (HFE) being retained across all three J-series missions. In addition to that, the ALSEP array consisted of a Lunar Surface Gravimeter (LSG), Lunar Ejecta and Meteorites (LEM) experiment, a Lunar Seismic Profiling (LSP) instrument and a Lunar Atmosphere Composition Experiment (LACE). Other experiments not part of the ALSEP array included a traverse gravimeter, a lunar neutron probe and a Surface Electrical Properties (SEP) experiment.

As with the two preceding J-series missions, the HFE experiment was designed to measure the temperature gradient over a depth of 96in (244cm). There was a degree of expectation over getting this to work, after the accident on Apollo 16 when the data cable was torn free of its connector to the Central Station, and there was a compromised set of data from the preceding flight as well. Changes had been made to the data cable for this HFE with a strain-release device added to remove its vulnerability to astronauts tripping over it!

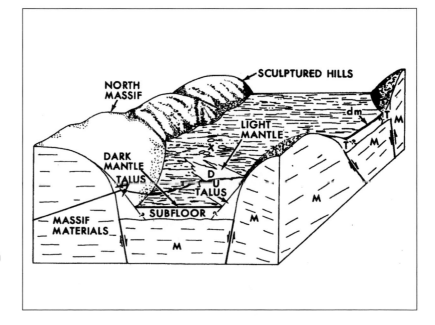

BELOW **The general characteristics of the Apollo 17 site on the eastern edge of Mare Serenitatis which, like Apollo 15's landing site on the eastern edge of Mare Imbrium, was nestled within a valley flanked by mountains.** *(NASA)*

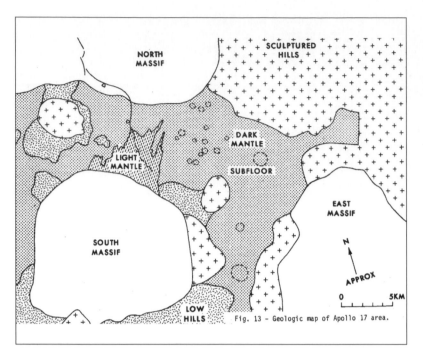

Fig. 13 - Geologic map of Apollo 17 area.

ABOVE The specific area investigated by this last lunar landing by humans for several decades to come presented opportunities to visit several geologic features including basaltic rock from the mountain-forming impact that created Mare Serenitatis as well as folded hummocks (the Sculptured Hills) and possible volcanic material on the valley floor. *(NASA)*

BELOW The basic scientific investigations distributed across all six lunar landings reflect the maturing technology made possible by an evolving succession of engineering changes made to the J-series hardware and the expanding contribution from the scientific community. *(NASA)*

EXPERIMENT	11	12	14	15	16	17
PASSIVE SEISMIC	X	X	X	X	X	
ACTIVE SEISMIC			X		X	
LUNAR SURFACE MAGNETOMETER		X		X	X	
SOLAR WIND SPECTROMETER		X		X		
SUPRATHERMAL ION DETECTOR		X	X	X		
HEAT FLOW				X	X	X
CHARGED PARTICLE LUNAR ENVIRONMENT			X			
COLD CATHODE IONIZATION		X	X	X		
LUNAR DUST DETECTOR		X	X	X		
LUNAR SURFACE GRAVIMETER						X
LUNAR EJECTA AND METEORITES						X
LUNAR SEISMIC PROFILING						X
LUNAR ATMOSPHERIC COMPOSITION						X
FAR UV CAMERA/SPECTROSCOPE					X	
LUNAR GEOLOGY INVESTIGATION	X	X	X	X	X	X
LASER RANGING RETRO-REFLECTOR	X		X	X		
SOLAR WIND COMPOSITION	X	X	X	X	X	
LUNAR SURFACE CLOSE-UP CAMERA	X	X	X			
COSMIC RAY DETECTOR					X	X
LUNAR PORTABLE MAGNETOMETER			X		X	
LUNAR TRAVERSE GRAVIMETER						X
SOIL MECHANICS			X	X	X	X
SURFACE ELECTRICAL PROPERTIES						X
LUNAR NEUTRON PROBE						X

The Lunar Surface Gravimeter was designed to detect gravitational deformities caused by tidal forces within the Moon stimulated by the alignment of the Earth and the Sun. It was also to measure 'free Moon' oscillations caused by gravitational radiation from cosmic sources. Ever since Dr Joseph Weber claimed to have detected such waves, physicists and astronomers enthused over the possibility of detecting these signs of violent activity deep in the Universe. Earth is too active to record the minute waves so the Moon is well placed to serve as a unique tuning fork.

The instrument consisted of a modified La Everte-Romberg gravimeter, comprising a mass suspended on a spring in a fused quartz glass frame. The suspended beam, pivoted at one end and connected to the spring, was balanced between two capacitor plates for electrical measurement of the characteristic vibration. Because of the requirement to constrain the thermal flux, a shield was attached to the upper portion of the box to prevent direct exposure to the Sun's rays. It weighed 28lb (12.7kg) and was to be set up approximately 25ft (7.7m) from the Central Station.

The Lunar Ejecta and Meteorites experiment was designed to provide data on primary and secondary particles striking the lunar surface by means of detector plates capable of measuring impact velocities from 0.6ml/sec to 46.6ml/sec (1km/sec to 75km/sec). The instrument contained an upper and a lower sensing platform. A dust particle falling to the surface at speed would penetrate a grid and pass through an upper sheet of film, travelling down to repeat the process on a second grid/film combination. An electrical pulse created by puncture of the film above and below displayed velocity and speed versus pulse strength and indicated mass. The exposed grid was divided into 256 sensors, all grouped into 16 basic sections.

The sensitive detectors were protected from dust and fine particles blown across the instrument by the Ascent Stage on lift-off by a cover jettisoned by ground command after the crew returned to lunar orbit. Set up 25ft (7.7m) from the Central Station, it would prove valuable in helping scientists study the amount of material deposited on the lunar surface when the Earth–Moon system passed through comet tails.

The Lunar Seismic Profiling experiment was essentially a data-gathering platform, the central module of which was placed on the lunar surface some 30ft (9.1m) from the Central Station. Two geophones (Nos 1 and 2) were set up in opposing directions, each 150ft (45.7m) from the module. Geophones 3 and 4 were laid out in a line at right angles to the first line, one 88ft (26.8m) from the module and the other in the opposite direction 260ft (79.2m) from the module.

This array established an equilateral triangle at the outermost geophones, with eight charges laid out at varying distances from 525ft (160m) to 1.49ml (2.4km), set down on the surface by the crew on the three EVAs. Each charge had a specific weight from 0.125lb (0.056kg) to 6lb (2.7kg), activated by three pins pulled at the time of deployment, timers delaying activation from 93hr to 96hr after deployment. A coded pulse triggered from the Central Station by ground command would detonate each cartridge charge, after which the geophones would continue to be used to measure local impacts from meteorites or from other impacts at more distant places as time went on.

The profiling equipment was more refined in its value compared with the Active Seismic Experiment, which involved geophones and a thumper device used on Apollos 14 and 16. The specific alignment of the geophone array, and the equilateral configuration of its placement, provided a numerically more valuable set of data points from Apollo 17. The particular location of each charge would be verified by the LRV's navigation equipment although specific accuracy was not critical to effective recording of the associated seismic waves.

The Lunar Atmosphere Composition Experiment (LACE) was designed to add further to an understanding of the Moon's 'atmosphere', which was transient and caused primarily by the gases such as carbon monoxide, hydrogen sulphide, ammonia, sulphur dioxide and water vapour released through volcanic activity, albeit in very small quantities. In fact the LACE instrument was equipped to measure hydrogen and helium as well at the bottom end of the atomic mass scale. Basically a Neir-type magnetic field spectrometer, the device collected gases through an inlet manifold,

passing them through an electron beam ionising filament before convergence in a focusing assembly and passage through to a magnetic field. Here, the circulatory motion of the particles enabled measurement of the quantity, lense (a transparent optical device used to diverge transmitted light) mass and the number of stripped electrons.

The spectrometer contained a layer of reflecting mirrors for thermal control and dust protection, removed by ground command after the crew left the surface, at which point the entire assembly would become active. The instrument scanned in three ranges of atomic mass units: 1–4, 12–48 and 40–110. After being placed on the surface approximately 45ft (13.7m) from the Central Station, a heater would be activated to drive out contaminating gases but would only be turned on after the Ascent Stage had left the surface.

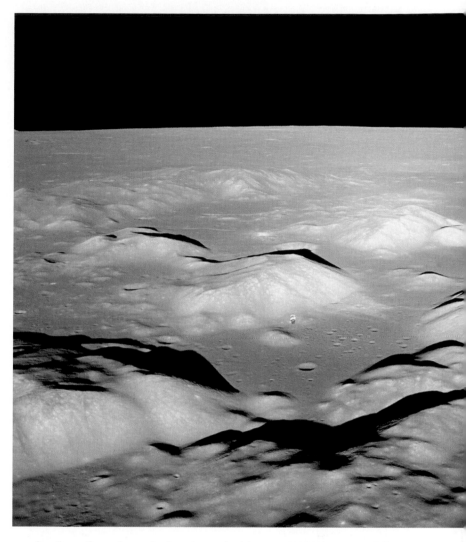

ABOVE A view of the Taurus-Littrow valley from the Lunar Module after separating from the Apollo spacecraft, which can be seen as a dot in the distance appearing down toward the surface but in fact in orbit. *(NASA)*

Not a part of the ALSEP array, the traverse gravimeter borrowed its concept from the highly successful tool employed by Earth-based geologists in determining internal structure, providing detail on the lateral distribution of dense materials. Mounted on the LRV, it would be used at the Lunar Module and then moved to remote locations with similar surface features to those on Earth for comparative analysis as well as for assisting in calibrating the difference between gravimetric measurements on each body. The instrument was sufficiently sensitive to measure the different gravity responses of craters, scarps and thickness variations in the overlying regolith.

The Surface Electrical Properties (SEP) experiment was designed to measure electrical properties at six selected frequencies from 1mHz to 32.1mHz, propagating signals from a self-contained transmitter with a multi-frequency antenna deployed on the surface approximately 300ft (91m) east of Challenger. Mounted on the LRV, the receiver would record reflected signals via a wide-band, tri-axis antenna with the precise location of the rover obtained from its navigation system. Final readings would be made at the end of EVA-3 and the recorder element returned to Earth. This experiment was integrated with data from the Traverse Gravimeter and from the seismic profiling experiment to build a consolidated and comparative data-set building a better picture of the Moon's outer layers.

The Lunar Neutron Probe was to be deployed in the bore hole of a deep drill core sample to determine the level of neutron flux in the upper layer of the surface. Some 7.8ft (2.4m) in length and with a diameter of 0.78in (2cm), the probe would be inserted in the hole during the first EVA and activated by a crewmember for collection at the end of the third EVA and returned to Earth.

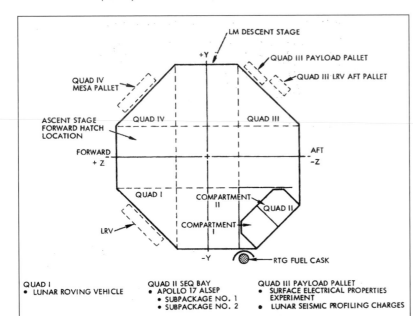

QUAD I
- LUNAR ROVING VEHICLE

QUAD II SEQ BAY
- APOLLO 17 ALSEP
  - SUBPACKAGE NO. 1
  - SUBPACKAGE NO. 2

QUAD III PAYLOAD PALLET
- SURFACE ELECTRICAL PROPERTIES EXPERIMENT
- LUNAR SEISMIC PROFILING CHARGES

QUAD IV MESA PALLET
- LUNAR SURFACE DRILL ASSEMBLY
- LUNAR NEUTRON PROBE
- LUNAR SURFACE ELECTRIC HASSELBLAD CAMERA
- CAMERA ACCESSORIES
- GROUND-COMMANDED TV CAMERA, CONTROL UNIT, AND ACCESSORIES
- SAMPLE RETURN CONTAINERS (2)
- LUNAR SURFACE RAKE
- SAMPLE CONTAINMENT BAGS (6)
- Sample Return Bag

QUAD III LRV AFT PALLET
- LARGE SAMPLING SCOOP
- GNOMON/COLOR PATCH
- HAMMER
- 32-INCH TONGS (2)
- TOOL EXTENSION HANDLE (2)
- LUNAR GRAVITY TRAVERSE EXPERIMENT
- LRV SAMPLER
- SAMPLE COLLECTION BAGS (2)
- EXTRA SAMPLE COLLECTION BAGS (4)

Night launch

The hardware for Apollo 17 began arriving more than two years before the launch date, including CSM-114, named America, and LM-12, to be called Challenger, with the last three-stage Saturn V to fly, AS-512, which was the heaviest of the vehicles in this class. Rollout to LC-39A took place on 28 August 1972 with several tests taking place in the 14 weeks to lift-off. These included an electrical test on 11 October and a three-day Flight Readiness Test that began on 18 October followed by CDDT on 20 November, the crew participating in the unfuelled 'dry' component of that activity.

The landing site for Apollo 17 dictated the launch window and the trajectory selected, which was a little different to earlier flights, a factor that determined the pane in that window. Launch time was set for 21.53hr on 6 December, opening a window that lasted 3hr 38min. As the first night launch in US manned space flight history, confidence trounced caution for crew recovery resulting from an abort – a requirement that kept earlier launches on the daylight side of the clock. NASA was excited about the fact that the giant Saturn V would light up the sky, visible in several States, the public relations machine pumping out advisories to news agencies proclaiming its visibility, somewhat doubtfully, halfway up the eastern seaboard.

Because of a malfunction in the automated sequencer the countdown went into a hold at T-30sec. Any delay would alter the shape of the outbound trajectory and in Houston flight controllers scurried to redefine the parameters for Trans-Lunar Injection. By reprogramming the guidance computer in the Saturn launch vehicle they could make up time during the trans-lunar flight so that Apollo 17 would arrive at the pre-planned time. This was important because all subsequent manoeuvres had been calculated from the time of Lunar Orbit Insertion.

After a delay of 2hr 40min, the only one in the Apollo programme caused by a hardware failure, AS-512 was launched at 00.33hr, 7 December, on a modified azimuth to an Earth orbit of 106.4ml x 104.9ml (171.2km x 168.8km). The launch azimuth was originally planned for 72° east of north but this was changed to 91.5° to shorten the transit time which was factored in to the guidance system

on the launch vehicle. The ability to do that and accommodate moderate delays had been built in to the crew preparation training and to the simulated operations regarding lighting angles and landmark tracking at the Moon.

The sequence of TLI closely followed that of previous lunar missions with the exception that, for the first time, the second burn of the S-IVB occurred over the Atlantic Ocean at the beginning of the third orbit instead of over the eastern Pacific toward the end of the second revolution. That burn began at 3hr 12min 37sec, exactly 39min later than the TLI burn for Apollo 16, and lasted 5min 51sec. Separation, transposition and docking to the LM was achieved by 3hr 56min 45sec, followed by ejection from the third stage 48min 15sec later. The S-IVB was targeted for an impact on the Moon, which occurred at 86hr 59min 41sec at 4.33°S by 8°N and a velocity of 8,346ft/sec (2,544m/sec), total mass at that time being 30,712lb (13,931kg).

The mass of the docked CSM/LM separated from the upper components of the Saturn launch vehicle was 103,167lb (46,796kg), compared with 103,175lb (46,800kg) for Apollo 16. Of the total, for Apollo 17 CSM-114 had a mass of 66,893lb (30,343kg) and LM-12 a mass of 36,274lb (16,453.8kg), the heaviest Lunar Module flown.

The trajectory selected invoked a longer transit time of 83hr versus less than 76hr for Apollo 16 and Apollo 15 and less than 72hr for Apollo 15. This was so that the pericynthion point could

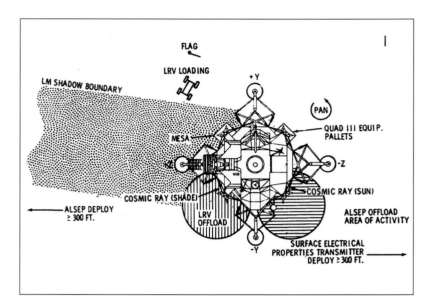

ABOVE The distribution of equipment and activity around the Lunar Module shows the shaded area and the various activity zones for unpacking the Descent Stage. *(NASA)*

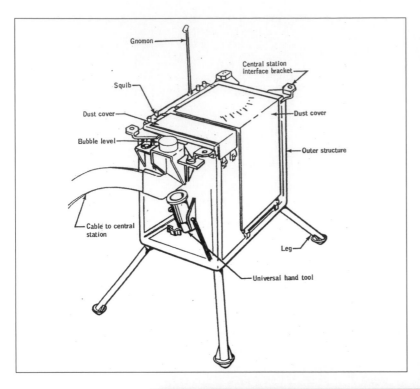

be relocated for a different set of orbits prior to landing and to reduced time spent in the Moon's shadow on shorter routes which could otherwise place critical demands on the thermal control system. However, had the original pre-launch trajectory been flown the cislunar cruise time would have been 2hr 40min longer. Nevertheless, the record of previous missions to the Moon was opening a greater flexibility and a new way of accessing difficult places.

The early part of the flight provided its fair share of concern. One of the docking latches failed to engage properly, a spectacular shower of frozen particles streamed past the Command Module windows, the caution and warning system buzzed and flashed incessantly and gastric upsets plagued Cernan. Several times during the outward flight the crew overslept. Music, voice calls, klaxons, sirens, bells and alarm tones were all employed before the sleeping trio could be awakened. But

ABOVE The Lunar Ejecta and Micrometeorites (LEAM) experiment with levelling device and data cable. Unique to Apollo 17, it was hoped the LEAM would help define the frequency of micrometeorite impact but some suspicious data cast doubt on its value. *(NASA)*

RIGHT The ALSEP site, with LEAM in the foreground, and the Central Station in the background and the RTG to the right. *(NASA)*

the mission did eventually settle down into unprecedented normality. However, approaching the Moon the indicated pressure in cryogenic hydrogen tanks 1 and 2 began fluctuating rapidly and for the remainder of the flight they were controlled manually.

In realigning the timeline of the Flight Plan, the hour between 46hr and 47hr GET was removed followed by removal of 1hr 40min at 66hr, rescheduling the onboard navigation tasks at 67hr 40min to align the schedule of activities with the compressed cislunar transit time. Mission Control had put forward its clocks in the MOCR room by 2hr 40min at 65hr true elapsed time. Notwithstanding some minor technical issues, one of which concerned the helium supply to the SPS engine, the flight to the Moon was uneventful and the planned checkout of the Lunar Module satisfactorily verified that all was as it should be in Challenger. A single mid-course correction of 9.9ft/sec (3m/sec) was

performed by the SPS at 35hr 30min in a burn lasting 1sec.

The special trajectory of Apollo 17 prevented it from entering a lunar shadow before orbit insertion but several hours before that a small crescent of the Moon became visible, growing rapidly until, at a distance of about 8,000ml (12,872km), features became visible as the docked vehicles swept around to the far side and a closest approach of about 60ml (96.5km).

Apollo 17 disappeared behind the limb of the Moon at 86hr 3min, just 10min before the LOI burn with the SPS which began at 86hr 14min 23sec and lasted 6min 33sec for a velocity change of 2,988ft/sec (910.7m/sec). Incorporated into this manoeuvre was a plane change of 11°, exceeding the magnitude of vectors incorporated into LOI on previous flights. When Apollo 17 came around the Moon into view of Earth, it was in an orbit of 195.6ml x 60.5ml (314.7km x 97.3km).

LEFT The Lunar Seismic Profiling (LSP) equipment is deployed with the antenna in the foreground for activating charges after the crew leave. *(NASA)*

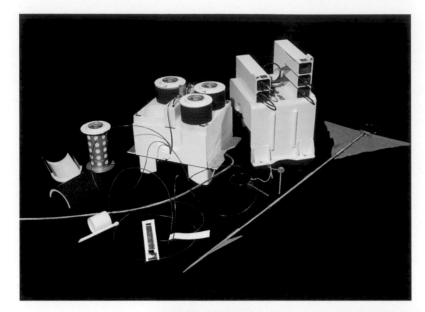

Activities between LOI and PDI, the latter planned for revolution 13, involved a significant change in operations. At the beginning of revolution 3, the first of two Descent Orbit Insertion burns (DOI-1) involved an SPS firing for 22.3sec and a ΔV of 197ft/sec (60m/sec) to reshape the orbit to one of 67.9ml x 16.68ml (109.2km x 26.8km). The two docked vehicles would remain in this orbit (modified to some extent by the mascons) until revolution 12 when Challenger would undock from America, the latter circularising its orbit about 90min later followed five minutes after that by Challenger doing the DOI-2 manoeuvre on the far side of the Moon and, coming around the eastern limb, begin powered descent to the surface.

The rationale behind all this was to effect a

higher altitude for Challenger coming around from the Moon's far side, providing an additional three minutes of ground tracking before PDI. Moreover, the DOI-1 burn placed the docked vehicles at a pericynthion altitude of 86,000ft (26,213m) compared with about 54,000ft (16,460m) for Apollo 16. The pericynthion was shifted to 10° west of the planned landing site versus 16° east. By splitting the DOI burn, and reverting to the Lunar Module responsibility for placing itself in the final elliptical orbit for PDI, it allowed for less effect from the mascons during the low pass, significantly reducing the probability of a wake-up call to lift pericynthion due to that gravitational torque on the trajectory.

The DOI-2 burn conducted by Challenger would lower pericynthion from 80,000ft (24,384m) to about 43,000ft (13,106m) for PDI, requiring only 40lb (18.1kg) of propellant for a net gain of three seconds in the hover toward the end of the landing phase. The net effect for spacecraft America would be to save 25ft/sec (7.6m/sec) of valuable SPS propellant, which was always felt desirable in the event that the Apollo CSM had to use its SPS engine to hunt down a disabled Lunar Module stranded in a low orbit after leaving the surface of the Moon.

Down to the Moon

And so, after Cernan and Schmitt settled in to Challenger, in accordance with that significantly revised and fine-tuned sequence, at 107hr 47min 56sec, the two spacecraft separated and Evans blipped the RCS thrusters for three seconds to effect a 1ft/sec (0.3m/sec) separation burn placing America in an orbit of 70.7ml x 13.2ml (113.7km x 21.2km).

For the entire front side pass of revolution 12, the two vehicles slowly drifted apart and 90min later, near the end of that orbit and on the far side of the Moon, when they were some 5,200ft (1,585m) apart, the SPS engine on America was fired at 109hr 17min 29sec for a fraction over 3sec for a velocity change of 70.5ft/sec, placing the CSM in an orbit of 80.5ml x 62.1ml (129.5km x 99.9km).

Five minutes later, at 109hr 22min 42sec, Challenger fired its RCS thrusters for 21.5sec, a ΔV of 7.5ft/sec (2.2m/sec), placing the LM in an orbit of 68.6ml x 7.1ml (110.3km x 11.4km). The two-stage hybrid manoeuvre had been performed without incident and for the first time the LM had only to use its RCS thrusters for the final orbit prior to descent. The gremlins of the outbound leg of the voyage had seemingly been left in cislunar space!

Hitting a record for the time taken to get a state vector update to Challenger's guidance computer after coming around on revolution 13, a mere 13 minutes before starting the descent, an update of 3,400ft (1,036m) was applied two minutes after ignition for PDI, which occurred at 110hr 09min 53sec. Coming on down at 30,000ft (9,144m), the high peaks of the South Massif became clearly visible and the divergence between the pre-planned trajectory and the on-board solutions converged. Landing radar velocity data came

BELOW The Lunar Surface Gravimeter (LSG) unpacked and placed on the surface at the ALSEP site. *(NASA)*

The landing site for Apollo 17 could not have been considered an early candidate for several reasons: it was, at 30.75°E farther east of the central meridian than any other mission, minimising the time taken to perform tracking and to generate updates for the Lunar Module prior to starting its descent to the surface; and at 20.15° it was at high latitude calling for a very special form of cislunar trajectory and for a major plane change on the way into lunar orbit.

But the potential science return from this site was immense, situated as it was within reach of several events on the stratigraphic column which shape this feature on the Moon today. But the initial choice for the last mission had been Alphonsus, a site much closer to the equator and west of the central meridian containing material from the lunar highlands and young volcanic material from the crater floor. As outlined earlier, the final selection came down to photographs returned by Worden on Apollo 15, a mission conducted at a latitude which allowed him to see that far north.

Paradoxically, it was believed that a considerable portion of the surface at this site was highland material uplifted when the basin was formed by a giant impact such as the one which had formed the Imbrium Basin. And the flat valley on which the landing site rested was thought to be covered with a fine-grained mantle consisting of volcanic ash materials.

The area was faced on three sides by steep mountains, massifs which probably were formed by breccias created by impacts that were of pre-Imbrian age and formed as well by the events that sculpted the Nectaris, Crisium and Serenitatis basins. If the majority of the massifs were laid out by the Serenitatis impact it was reasoned that the highland crust would lie within inches of the fine particulate accumulations.

To the north-east of the site lay the Sculptured Hills which were thought to be of the same origin as the massifs but several selenologists believed them to have some form of volcanic origin. Across the plain itself upon which Challenger would land, no large blocks and boulders were present. A major feature at the site was a 260ft (80m) scarp named Lincoln-Lee. Overlaying this was a bright mantle with finger-like rays that appear as a small avalanche off the slopes of the south massif, as if dislodged by a seismic event.

RIGHT This shows the basic components of the LSG, a flawed instrument which suffered from some intrinsic errors in its design and assembly. *(NASA)*

in at 3min 35sec followed by radar range-to-go data 38sec later.

At 13,000ft (3,962m), Cernan could clearly see the mountains out of his window and as throttle-down occurred and High Gate was reached at 7,500ft (2,286m), 9min 22sec into the burn, it was apparent that there were better places to touch down than the designated landing spot indicated by the LPD. Cernan took over manual control at 300ft (91.4m) and 'flew' Challenger further south to avoid large blocks and craters which had not been apparent in the best orbital photography. On touchdown at 12min 05sec (110hr 21min 58sec) after a flawless descent, the LM rocked back as the aft leg sank into a depression which the crew observed later may have 'stroked' the compressible shock absorption material in the leg, something not seen on previous landings.

The descent had been a special event

for Cernan – the second time he had been down close to the Moon. On Apollo 10, he had simulated all the events up to, but not including, ignition of the LM's propulsion system for landing. That flight had cleared the way for the final test of Apollo, the actual landing, on Apollo 11; now, he had been back to conduct the segment that had eluded him before. His ebullience on landing said it all: 'OK Houston, the Challenger has landed…Boy, when you said "shut down", I shut down and we dropped, didn't we…Yes sir, but we is here, man is we here! Houston, you can tell America that Challenger is at Taurus-Littrow.'

After the customary confirmation from Mission Control that they could remain on the surface, Cernan and Schmitt began a methodical and systematic preparation for their first excursion but by the time they were ready to leave Challenger they were already almost an hour behind schedule. Each crew is different in the way they proceed with these preparations and Cernan executed a modified flow of activity which compromised the sequence – to his credit because it created an orderly space inside Challenger and made neatness a virtue; but all crews are different in those activities.

The first EVA began at 114hr 21min 49sec and Cernan commented on the extreme blackness of the porch area where he crawled backwards, on all fours, to emerge and descend the ladder, advising Schmitt 'not to lower your gold (coated) visor until you get on the porch because it's plenty dark out here'. Within 30min the LRV was offloaded. Then it was fully deployed and a test drive performed and photographed with some local samples gathered and documented. The 500mm photographic panoramic survey was conducted and at 1hr 21min the US flag was set up with some posed shots of each astronaut.

Over the six landings, the way in which photography was exploited had changed dramatically. The almost casual afterthought given to the tasking on Apollo 11 left only a couple of pictures of Neil Armstrong; it was the Commander whose job it was to do most of the picture-taking of activities and procedures showing Buzz Aldrin, who never specifically took a picture of Armstrong for history. The J-series missions had changed the entire

ABOVE The Lunar Atmosphere and Composition Experiment (LACE) which furthered research into the atmosphere of the Moon, specifically into hydrogen and helium present around the lunar sphere. *(NASA)*

BELOW The ALSEP array and orientation of equipment and experiments, with the geophone array for seismic profiling and the relative location of the two Heat Flow Experiment drill holes. *(NASA)*

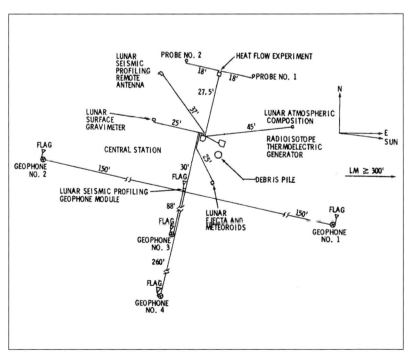

process by which the missions, the geological expeditions and the attention to science as well as the historical nature of the activity reached much higher levels of priority – it was as though the uniqueness of these few landings began, finally, to take hold.

For this mission a TV camera failed to show the initial activities, the timeline being too packed to follow the precedents set by Apollo 11, but the energy with which they thrust themselves into ALSEP extraction, offloading experiments and the start to deployment caused excessive energy levels that brought a request from Mission Control for them to slow down. The timeline for EVA-1 had already been loaded – incorporating the Heat Flow Experiment had not originally been planned for Apollo 17 but with the loss of that experiment on Apollo 16 it was included in the already crowded lunar surface operations plan for

this last mission. It had been a much troubled experiment, having first been flown on Apollo 13 (which of course never landed) and again on Apollo 15 where the deployment was partial and compromised by a less than effectively drilled hole.

While Schmitt set about deploying the ALSEP experiments, Cernan began drilling for the heat probes, experiencing difficulty drilling through the regolith and Schmitt was required to go over and apply some muscle extracting the stems by physically throwing his entire weight on to the jack. Several times during the ALSEP deployment the crew used the traverse gravimeter to obtain readings but did manage to obtain a deep-core sample into which the lunar neutron probe was emplaced. Although the deployment of ALSEP instruments went according to plan, there were several time-consuming issues which delayed progress beyond the rigours of the scheduled timeline.

Because the ALSEP took longer than expected, the two lunarnauts were about 40min behind the timeline when they departed on their first geology trip at 4hr 50min. Because of this, instead of going to a crater called Emory (station 1) some 1.49ml (2.4km) to the south-east as planned they were directed to a position exactly half that distance out from Challenger at a crater called Stenor, passing a set of craters called Triplet on the way. The geological objectives set for Emory would be satisfied at Stenor.

Designated station 1a, they got to Stenor at 5hr 2min after a 13min drive over an undulating surface, where they deployed one of the seismic profiling charges, obtained a further traverse gravimeter reading and conducted documented rake samples before an extensive set of panoramic photographs. Leaving station 1a at 5hr 35min, they drove back to the site quite close to the Lunar Module where the Surface Electrical Properties experiment was to be set out, arriving there just 14min later. They deployed the two antennas across the lunar surface in opposing directions spanning 230ft

OPPOSITE The deployment sequence for the Lunar Roving Vehicle, applicable to all three J-series missions. (NASA)

BELOW Inserted into a bore hole, the Lunar Neutron Probe showing length and separate components of the instrument and its detectors. (NASA)

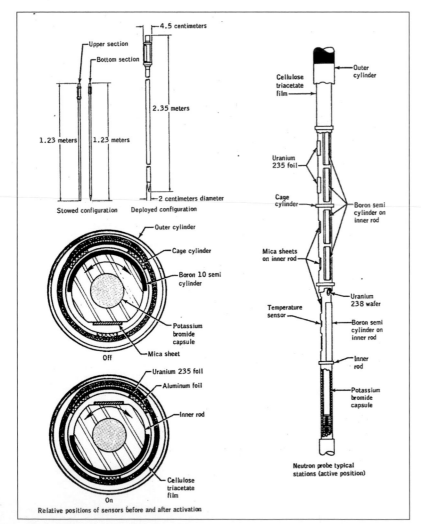

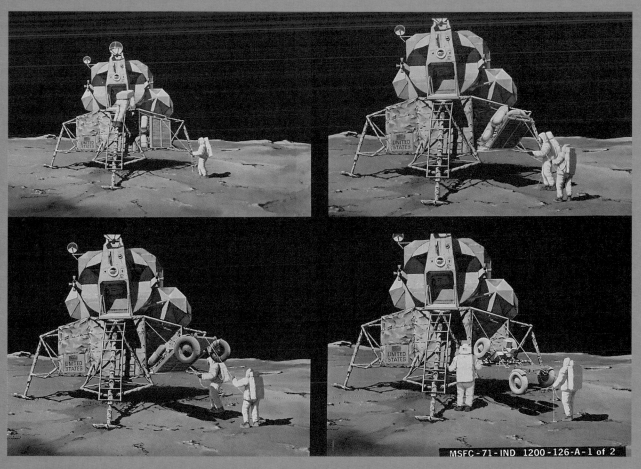

MSFC-71-IND 1200-126-A-1 of 2

MSFC-71-IND 1200-126-A-2 of 2

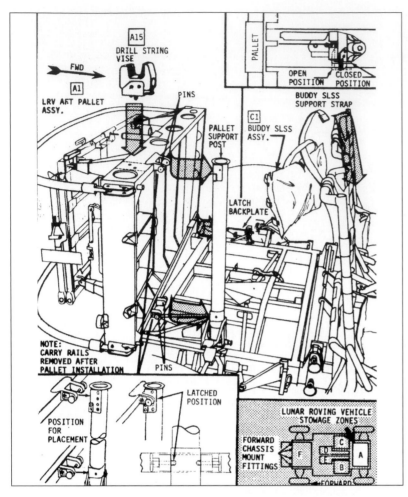

LEFT, BELOW AND OPPOSITE PAGE The astronaut orientation manual for the Lunar Roving Vehicle shows equipment attached to various parts of the structure, as identified by the placement diagram at bottom right in each illustration. (NASA)

(70m) together with the transmitter, ready for the second and third EVAs. Cernan and Schmitt spent the remainder of the time there on another gravimeter reading, gathering samples, conducting documentary photography and taking panoramic pictures.

Leaving the SEP site at 6hr 12min into EVA-1, they took the LRV a short three-minute drive to Challenger and began closeout activities which included two more gravimeter readings for a total of nine. The EVA ended with depressurisation at 7hr 11min 53sec, a mission elapsed time of 121hr 33min 42sec. Pre-flight timelines allocated nearly 1hr 10min at station 1, abbreviated to little more than 30min due in part to the late start and also to the additional time taken on ALSEP deployment. Nevertheless, the LRV logged approximately 2.17ml (3.5km).

The Flight Plan had them going to sleep at 125hr 20min, almost exactly 15hr after landing, but an intensive debate over the geology at the site and about their observations during the first EVA kept them occupied for some time, the rest period eventually beginning at around 126hr 15min. The call-up from Mission Control went in at 134hr 15min with Cernan and Schmitt already up and eating before beginning preparations for EVA-2.

From valley floor to the hills

EVA-2 began at 137hr 55min 06sec, some 27hr 33min since touchdown, but more than 1.5hr later than scheduled, the first hour being spent preparing the LRV with tools and equipment and some additional time spent repairing the rear fender which had broken on EVA-1. They drove to the SEP site at 49min and arrived three minutes later, where they activated the experiment and left for station 2 at 56min elapsed time, stopping at four short intervals, one to deploy another seismic profiling charge and three for quick sample collection.

The boulder population increased

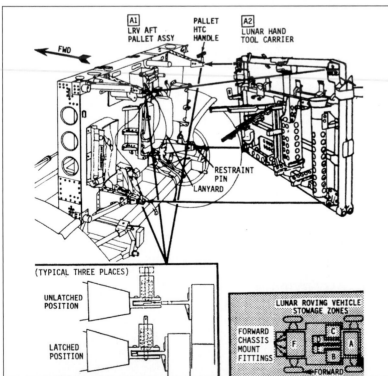

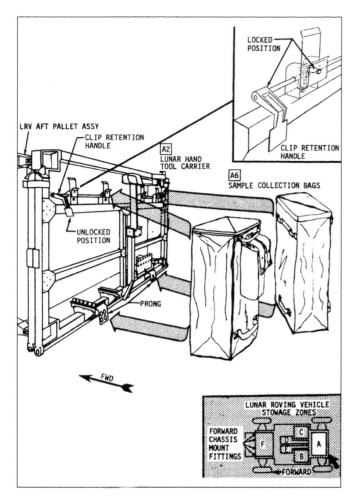

LRV AFT PALLET ASSY
CLIP RETENTION HANDLE
A2 LUNAR HAND TOOL CARRIER
A6 SAMPLE COLLECTION BAGS
LOCKED POSITION
CLIP RETENTION HANDLE
UNLOCKED POSITION
PRONG
FWD

LUNAR ROVING VEHICLE STOWAGE ZONES
FORWARD CHASSIS MOUNT FITTINGS
FORWARD

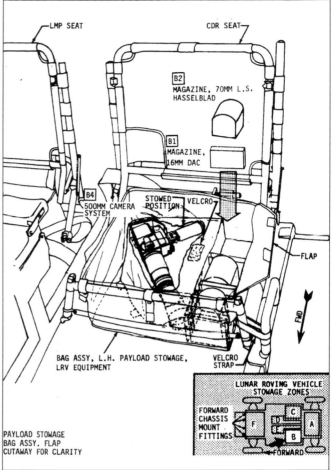

LMP SEAT
CDR SEAT
B2 MAGAZINE, 70MM L.S. HASSELBLAD
B1 MAGAZINE, 16MM DAC
B4 500MM CAMERA SYSTEM
STOWED POSITION
VELCRO
FLAP
FWD
BAG ASSY, L.H. PAYLOAD STOWAGE, LRV EQUIPMENT
VELCRO STRAP

PAYLOAD STOWAGE BAG ASSY. FLAP CUTAWAY FOR CLARITY

LUNAR ROVING VEHICLE STOWAGE ZONES
FORWARD CHASSIS MOUNT FITTINGS
FORWARD

dramatically as the LRV wheeled its way across the light mantle material, swept from the slopes of South Massif onto the floor of the valley. More than half a mile (a kilometre) along, the rover encountered rough terrain. Soon they came up on Horatio, another crater to their right, as they moved almost due west, 1.2ml (2km) from the Lunar Module. The scarp was not too evident as they headed directly for it; if they approached it, they would approach it on the down-fault side, appearing to them as a near-vertical cliff rising in places as high as 262ft (80m) from the valley floor. But they were now heading too far south to encounter Lincoln scarp and, as planned, would sample the light mantle where it met the valley floor. Then they moved back on to darker material after stopping for another quick grab-sample.

They were closer to the mountain now and starting to move up the undulating slopes. 'Yeah, there's no question that there is apparent lineations all over these massifs in a

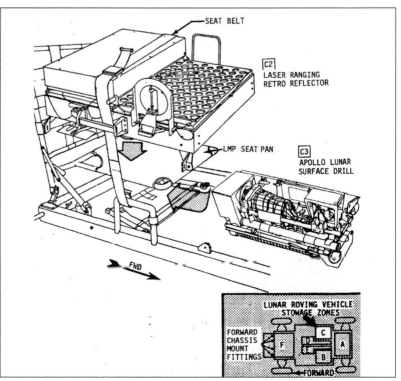

SEAT BELT
C2 LASER RANGING RETRO REFLECTOR
LMP SEAT PAN
C3 APOLLO LUNAR SURFACE DRILL
FWD

LUNAR ROVING VEHICLE STOWAGE ZONES
FORWARD CHASSIS MOUNT FITTINGS
FORWARD

variety of directions. Hey look at how that scarp goes up there,' cried Schmitt as Cernan controlled the rover.

'It looks like the scarp overlays the North Massif, doesn't it?' opined Cernan. They drove on past Lara, a crater seen in photographs from orbit and then by Nansen to stop on its flank at the spot called station 2: 'Boy, you're looking right into Nansen. We're right where we wanted to be for station 2. And it looks like a great place. Big blocks. It looks like quite a bit of variety from here. Different colours anyway, greys and lighter coloured tans.'

Finally, after driving for 80min they had reached their destination about 4.7ml (7.6km) from Challenger, but only after an intermediate stop at a freshly designated station 2a to get a reading from the traverse gravimeter which had been causing problems. Work at this and successive sites was highly productive due to the geological significance of the area with sampled blocks which originated on the slopes of the South Massif. The place was so productive that Mission Control told Cernan and Schmitt to remain an additional ten minutes beyond the pre-assigned time and take that out of a later stop. Busy with detailed activity they sampled the regolith, collected rocks, reported readings from the traverse gravimeter, raked loose material and retrieved rock chips.

The crater itself had been named after the Norwegian arctic explorer Fridtjof Nansen. 'When you look down into the bottom of Nansen it looks like some of the debris there has rolled off the South Massif and covered up the original material,' commented Schmitt, adding that 'all the boulders that have come down are on the south side of the slope'. They spent 1hr 6min at station 2 and departed at 3hr 12min, driving back to station 3 now 3.7ml (6km) from Challenger, arriving there 41min later.

Station 3 was located near the base of the scarp about 150ft (50m) east of Lara Crater and most samples here were collected from the rim of a crater about 30ft (9.1m) across but the geological attractiveness of

BELOW The complement of tools and essential geological sampling equipment attached to the LRV for a full traverse. *(NASA)*

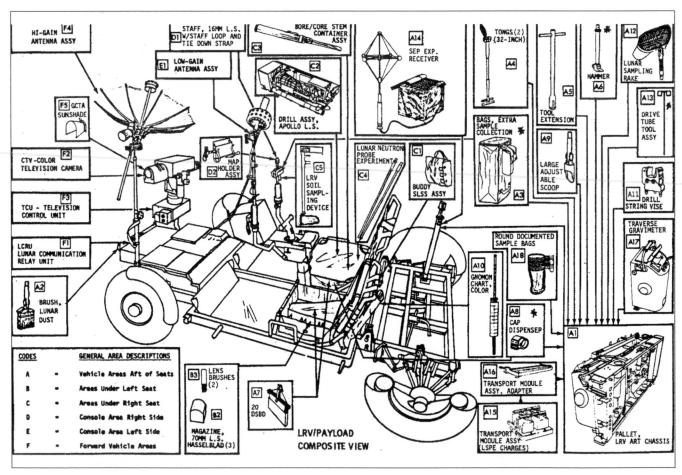

the site occupied them for 37min before they departed, urged along by Mission Control. But they were able to confirm that Lara Crater itself, about 1,650ft (500m) in diameter, was likely covered in the light mantle from South Massif. Because of this the origin of materials collected were probably a mix of material produced by the impact event that created Lara and from the mantling.

They left Lara at 4hr 30min and stopped briefly on the way to station 4, the first about 2,625ft (800m) north-east of site 3 where they believed the adjacent crater may have penetrated the light mantle into bedrock, and the second about 0.8ml (1.3km) north-east of station 3 and on the light mantle itself. The explorers reached station 4 at 4hr 48min where they would spend the next 36min sampling Shorty Crater, a depression with a diameter of about 361ft (110m) with a hummocky floor and with a distinctly raised rim featuring a central mound containing blocky rocks and boulders.

The crew took samples from a portion of the rim crest and south of a large 16ft (5m) boulder of fractured basalt which was itself surrounded by debris and rocky fragments, some buried, others only partially so. A trench was dug in the rim crest where Cernan discovered reddish soil in a zone about 3ft (1m) trending parallel to the crater rim crest for about 6ft (2m). Yellowish bands 4in (10cm) wide formed margins of the deposit and the astronauts saw other bands of a similar colour on the flanks of the crater. The most immediate assumption was that they had found volcanic material and the excited tones of both Cernan and Schmitt testified to their belief that one of the reasons for selecting this site had been vindicated.

It is unlikely that volcanism was the reason for the reddish soil, Shorty Crater itself having clearly been of an impact origin. But geologists concede that it may have been excavated from an underlying deposit and brought to the surface by that event. Or, a fissure may have run along the crater rim and produced an outflow of limited volcanic material, but its discovery was one of the seminal moments in the story of lunar exploration and the jury is still out as to exactly what this soil was and where it originated. But it certainly raised a clamour for answers!

A traverse gravimeter reading, trench sample and double-core tube sample together with documented and panoramic photography preceded the departure for station 5 at 5hr 24min, with a stop to deploy another seismic profiling charge and another to conduct some sampling, reaching their destination at 5hr 50min. Located in a blocky field close to the south-east rim of the crater Camelot, a 1,970ft (600m) diameter depression, station 5 consisted of a wide variety of blocky materials covering about one-third of the surface area. Young craters and depressions littered the entire region and were observed to be superposed on the rim of Camelot itself. Scientists hoped that the materials collected should have come from as deep as almost 100ft (30m), excavated by the impact that formed this crater.

Cernan and Schmitt left station 5 at 6hr 20min and headed back to Challenger, stopping briefly to deploy yet another seismic package, document that with photographs and attend to a few remaining items, including re-levelling the Lunar Surface Gravimeter instrument. They were back at the Lunar Module at 6hr 37min and commencing to close out EVA-2 after a traverse gravimeter reading at 7hr 24min. The

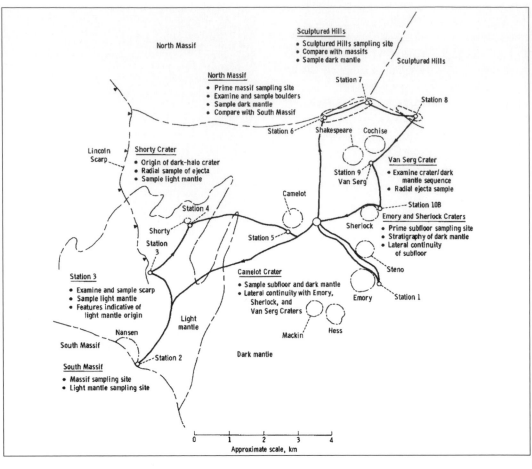

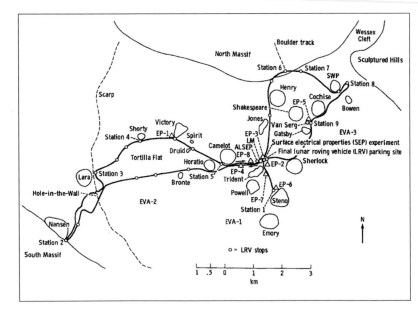

EVA ended at 7hr 36min 56sec, a surface time of 35hr 10min 05sec, Cernan and Schmitt having logged 12.67ml (20.7km) on the rover.

After attending to the spacecraft's cabin pressure regulator – Mission Control thought there was a slight leak – the crew recharged their backpacks and had a well-deserved meal. Ron Evans in America had been asleep for some time and now it was time for Cernan and Schmitt to begin their last 'night' on the Moon. Despite a late start and an extended EVA, they were only one hour behind schedule when Mission Control said goodnight, at 149hr 45min mission elapsed time, 39hr 23min after landing.

As a result of continued concerns about the amount of work expended and the sustained exertions of working on the lunar surface, the physicians were adamant that the crew needed a full 8hr rest period and the wake-up call went up to Challenger at 157hr 45min, 47hr 23min into their surface time and about 28hr before they would leave for a rendezvous and docking with mothership America. But they had slept only fitfully, Cernan getting only three hours' deep sleep. With each passing day on the Moon the Sun was rising higher, the Sun being 15° above the horizon when they began EVA-1, 25° at the start of the second and now, for the third EVA, 40°, its dominating presence felt in the higher temperatures.

The last Moonride

EVA-3 began at 160hr 52min 48sec and shortly after reaching the surface the two explorers conducted another traverse gravimeter reading, loaded LRV-3 for a full geological field trip, retrieved the cosmic ray experiment and set off for the SEP site at 44min elapsed time, arriving a couple of minutes later. There they activated the experiment, gathered samples and conducted a documented photographic survey, departing for station 6 at one hour into EVA-3.

They were to move almost due north where, at the edge of the North Massif, they could sample large boulders dislodged millions of years ago by some seismic event. Moving around the rim of a crater called Henry, Cernan and Schmitt bounced and rocked along to the base of the mountain; having sampled the South Massif the day before, they would now visit the more spectacular scenery to the north end of the valley. From Lunar Orbiter photographs taken several years earlier, and in pictures returned by Apollo 15, geologists knew there were interesting boulders to sample at this location.

But when the crew approached it across the bumpy valley, it was quite spectacular. 'Oh man, what a slope,' called Cernan as he struggled to find a path. The boulders they wanted to see lay 262ft (80m) up the South Massif on a gentle incline of about 11° from which snaking tracks led higher up the mountain tracing a furrow gouged out as the rocks had rolled down. The sharpness of some boulder tracks indicated that they had rolled downhill only recently.

Stopping on the way for more samples, the second stop 1.7ml (2.8km) from the SEP site, they arrived at their first major sampling site at 1hr 19min where they would remain for traverse gravimeter readings, a single core sample and a rake sample with documentary, panoramic and 500mm photography. Station 6 was on the south slope of the North Massif approximately 656ft (200m) north of the line where the flat plain gives way to the hills.

They had stopped alongside a massive boulder 59ft (18m) long and 32ft (10m) wide by 20ft (6m) high, shattered now into five separate segments. It was a geologist's dream. From this vantage point above the lava bed of the plain across which they had driven stood their spacecraft more than 1.8ml (3km) away. The

LEFT A pre-flight view of the Traverse Gravimeter, not part of the ALSEP array but carried along on the LRV as the crew progressed along their geological traverse routes, making frequent measurements as they went. (NASA)

view was a montage of undulating hills, the bleak face of South Massif and rippled humps across the flat valley. The Sun was noticeably higher this day, equal to mid-morning on a summer day back on Earth.

But the lonely, desolate lunarscape made it unique and alluringly attractive, tranquil and utterly silent – even to the helmeted explorers to whom the only sounds were the very faint 'hiss' from the oxygen flowing into their suits and the engaging voice, sometimes a crackle, as they retained contact with Earth through their communications unit on the LRV. Even within the time-cramped urgency of routine tasks, meticulously rehearsed back on Earth, there were fleeting seconds to stand and gaze at the vista before them, made all the more beautiful by their elevated position up the slope of South Massif, and briefly savour the moment, recognising that they were the last of 12 men who would walk on the Moon for a very, very long time to come.

One of the boulders had a prominent north-facing overhang shading soil, two samples of which were taken in addition, one in sunlight. In addition, some soil was collected from a niche in one of the boulders, material possibly collected as it rolled downhill. In addition, rake samples were obtained as well as a single core tube within a very short distance of this boulder. Some considerable attention was given to the five segments of the large boulder displaying at least two major rock types which were a highly vesicular light-grey breccia and a darker blue breccia.

Cernan and Schmitt remained at station 6 for 1hr 11min working intensively on these tasks before leaving for station 7 at an elapsed time of 2hr 30min, which was a mere seven minutes away. This place was similar to station 6, located at the base of the North Massif and on a slope of 9° and similarly littered with boulders which had rolled down toward the valley floor. They sampled a 10ft (3m) boulder which appeared to be composed of complex breccia

elements and several chippings were collected here, the explorers noting the prolific abundance of tiny fragments and small blocks randomly strewn as though tumbling down from shattered boulders disintegrating as they rolled to a stop.

The soil at station 7 was fairly compacted, rover tracks and boots penetrating only one-half to one inch (1.2cm to 2.5cm) although only a short distance away the regolith was deeper. This location kept the astronauts occupied for 22min before they moved on to station 8, 16min further along, although they stopped briefly for another collection of material. Arriving at station 8 some 3hr 15min into the EVA and now 2.5ml (4km) from Challenger, they conducted trench sampling, took regolith samples, obtained soil from beneath a boulder, collected rock fragments and rock fines, obtained core samples and took a traverse gravimeter reading. They got a lot done in the 48min they were there before departing for the next site at 4hr 3min.

Now it was back down into the valley for station 9 and a crater called Van Serg, 295ft (90m) across and with a blocky central mound 98ft (30m) wide. Arriving at 4hr 21min, Cernan and Schmitt reported it to have a distinctly raised and blocky rim with an ejecta blanket defined by rocks and small boulders. The explorers focused their work at two locations, on the south-east rim crest and on the ejecta blanket itself. The explorers deployed another geophone charge, took two traverse gravimeter readings, gathered a trench sample and a double-core tube sample, conducted documented and panoramic photography and used their 500mm for distant views.

The great value in this was that as the Sun rose higher the lighting angles made a significant difference to the features visible on the massifs and on the Sculptured Hills. They were nearly an hour at Van Serg, departing there at 5hr 17min for a drive back to the Lunar Module. They stopped to gather samples and to deploy another seismic profiling charge with more panoramic and documented photography. The latter was valuable, the absence of a light-scattering atmosphere bringing into sharp relief any deformation or lineation, providing geologists with a rich set of contrasts and views that helped add dimension to the whole area.

BELOW Apollo 17 was the only lunar mission to carry the Traverse Gravimeter, which proved useful in defining the substructure at the Taurus-Littrow site. *(NASA)*

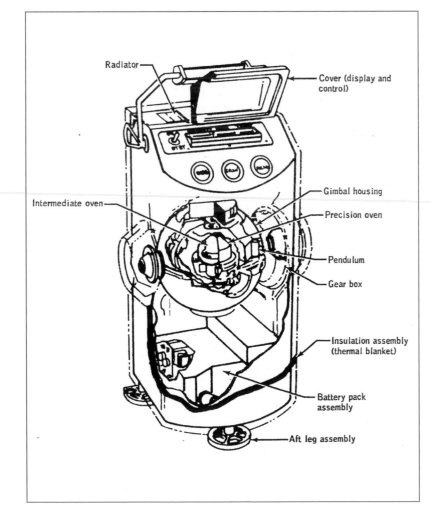

- Radiator
- Cover (display and control)
- Intermediate oven
- Gimbal housing
- Precision oven
- Pendulum
- Gear box
- Insulation assembly (thermal blanket)
- Battery pack assembly
- Aft leg assembly

LEFT A determined Jack Schmitt sets off at a pace to station 5, looking east. *(NASA)*

BELOW Given scale by the rover at right, the large rock at station 6 up the North Massif looking south. *(NASA)*

BOTTOM Schmitt works a massive boulder at station 6. *(NASA)*

Cernan and Schmitt were back at Challenger at 5hr 46min and beginning the closeout of the third EVA, arguably the most productive and scientifically rewarding mission of the three J-series expeditions. Two more traverse gravimeter readings were taken, the ALSEP photography was completed to provide verification of its integrity and the precise relationship of the various experiments to other items such as cover, protectors, lanyards, tethers and straps that littered the area.

At 6hr 16min the NASA Administrator Dr James C. Fletcher spoke to the crew and congratulated them on their success. Responding, Cernan reminded those listening that 'this valley of history has seen mankind complete its first evolutionary steps into the universe, leaving the planet Earth'. He concluded: 'I can think of no more significant contribution has Apollo made to history. It's not often that you can foretell history but I think we can in this case. And I think everybody ought to feel very proud of that fact.'

Only a few more tasks remained, including retrieval of the neutron probe at 6hr 44min after 49hr of exposure since it had been inserted into the deep core on the first EVA. Then at 6hr 48min, LRV-3 was driven to its parked position so that it could face Challenger and broadcast to Earth the ascent of the last Lunar Module to depart the lunar surface. As Cernan was about to jump up to the first rung on the ladder he had a few final words for the world: 'As I take man's last steps from the surface, back home for some time to come, but we believe not too long into the future, I'd like to just list what I believe history will record – that America's challenge of today has forged man's destiny of tomorrow. And as we leave the Moon at Taurus-Littrow, we leave as we came and, God willing as we shall return, with peace and hope for all mankind. God speed the crew of Apollo 17.'

The EVA officially ended with re-pressurisation at an elapsed time of 7hr 15min 08sec, a mission time of 168hr 07min 56sec. Much work remained to be done including preparing for a fourth opening of the hatch to jettison unwanted equipment and anything that would lighten the Ascent Stage for its departure. Then it was time for a meal and for an extensive discussion with Mission Control about their last geological field trip, details still fresh in their minds. And then, at 172hr 40min it was time for a very well-earned rest after a day that had lasted almost exactly 15hr since their wake-up call.

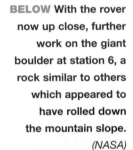

BELOW With the rover now up close, further work on the giant boulder at station 6, a rock similar to others which appeared to have rolled down the mountain slope. *(NASA)*

The timeline had been well adhered to and the physiological responses of the crew had been better than those on the two preceding missions, their LRV having taken them a distance of 7.5ml (12.1km).

Of great interest to both engineers and physicians was the performance of the crew in task-management and in the ability of the PLSS units to handle metabolic demands placed on the equipment by high workloads presented by each astronaut. In general, and despite some heavy work, Cernan consumed less oxygen, at 4.21lb (1.9kg) cumulative on all EVAs, versus Schmitt's consumption of 4.36lb (1.97kg). Battery power consumption was similarly biased toward a great quantity used by Schmitt, accounted for by higher cooling switch selections to offset a greater metabolic output.

Lessons for the future included time to clean down the superficial dust and fine particles which had, on all six lunar expeditions, been troublesome, as well as potentially dangerous, in that grainy material clogged connectors and inhibited a proper seal which on occasion, as indicated in this book, produced a minor leak in suit pressure. The closeout activity for EVA-3 was extended to give Cernan and Schmitt time to clean down as far as possible.

The last ride home

To the musical strains of Richard Strauss's *Thus Spoke Zarathustra*, the wake-up call from Mission Control flowed into the Lunar Module Challenger at 180hr 50min. Time to get up and time to leave the Moon. But as they slumbered on the lunar surface, above the Moon Ron Evans had performed an orbit trim burn with the RCS thrusters at 178hr 54min 05sec, a 31sec burn with a ΔV of 9.2ft/sec (2.8m/sec) which placed America in an orbit of 77.4ml x 71.9ml (124.5km x 115.7km).

About one hour later, so as to align the ground-track inclination with that of the Ascent Stage as it came up from the surface, Evans had commanded a plane change at 179hr 53min 54sec, a 20sec firing of the SPS engine for a ΔV of 366ft/sec (111.5m/sec) which adjusted the orbit to 72.2ml x 71.9ml (116.1km x 115.7km). There were still many preparations to make before Cernan and Schmitt could leave and the condition of the interior of Challenger

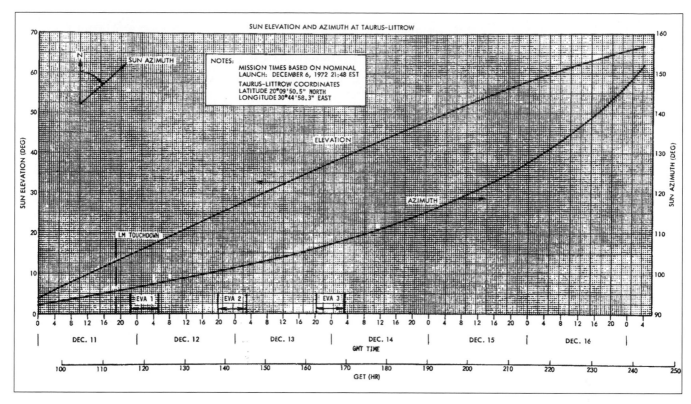

SUN ELEVATION AND AZIMUTH AT TAURUS-LITTROW

NOTES:
MISSION TIMES BASED ON NOMINAL
LAUNCH: DECEMBER 6, 1972 21:48 EST

TAURUS-LITTROW COORDINATES
LATITUDE 20°09'50.5" NORTH
LONGITUDE 30°44'58.3" EAST

was not a little compromised by the prolific quantity of dust that had settled on everything.

Lift-off occurred at 185hr 21min 37sec after a surface stay time of 74hr 59min 40sec, the Ascent Propulsion System firing for 7min 21sec, producing a ΔV of 6,075ft/sec (1,852m/sec) and carrying Cernan and Schmitt to an orbit of 55.8ml x 10.4ml (89.8km x 16.7km). A small vernier adjustment was required, performed at 185hr 32min 12sec, a 10sec RCS thruster firing for a ΔV of 10ft/sec (3m/sec), raising pericynthion to 10.8ml (17.4km) on the way to a first-orbit rendezvous. Terminal Phase Initiation occurred at 186hr 15min 58sec, a 3.2sec blip with the Ascent engine and a ΔV of 53.8ft/sec (16.4m/sec) conducted at apocynthion, raising pericynthion to 74.4ml (119.7km) and swapping that for a new apocynthion.

At acquisition of signal on revolution 52 the two spacecraft were less than 1.2ml (2km) apart. Before docking, Challenger flew around America inspecting the SIM bay to check out its condition and that of the deployments. Due to a very slow closure rate of less than 0.1ft/sec (0.03m/sec) docking was not achieved at the first try and a second attempt, at a faster rate, was necessary before the three small capture latches snagged the Ascent

ABOVE As the three EVAs progressed the Sun elevation angle rose appreciably, as shown by elevation and azimuth on this chart. *(NASA)*

BELOW Schmitt at work at Van Serg Crater, station 9. *(NASA)*

Stage drogue. As Challenger pitched down to get in the correct attitude, up across the forward windows came a view of the Taurus-Littrow landing site directly below. Capture was achieved at 187hr 37min 15sec, a total separation time of 87hr 15min 18sec.

Some considerable time was spent vacuuming the interior of the Lunar Module – in fact it was left on all the time the hatches were open between the two vehicles – and the Command Module was remarkably dust-free when the crew finally began to seal up the Ascent Stage for separation. It also helped greatly that Cernan and Schmitt were at pains

to bag everything carefully and to clean the sample return containers before shifting them across. A lot learned from preceding missions was coming together and learning curves were converging.

Jettisoning of the Ascent Stage occurred at 191hr 18min 31sec and the CSM fired its thrusters for 12sec precisely 5min later, imparting a ΔV of 2ft/sec (0.6m/sec) to separate the two vehicles, placing America in a very slightly modified orbit of 73.5ml x 70.4ml (118.2km x 113.2km). Unlike Apollo 16, the stage exited in a tidy and controlled fashion, its guidance system holding a tight attitude bandwidth for the de-orbit burn. That required a long 1min 56sec firing of the RCS thrusters which began at 192hr 58min 14sec, producing a decelerating ΔV of 286ft/sec (87m/sec). It struck the South Massif 19min 07sec later just 6.1ml (9.9km) from the Apollo 17 ALSEP array and only 1.1ml (1.75km) from the aim point.

The crew began their sleep period at about 196hr. It had been a long day, 16 hours in all, but a reasonably balanced work/rest cycle in which all the lessons of the previous flights provided a better fitness balance for the crew. In so many respects, Apollo 17 was the go-to template for future expeditions to the lunar surface. Lessons had been learned and the physicians had the measure of what was possible and what was wise and prudent to

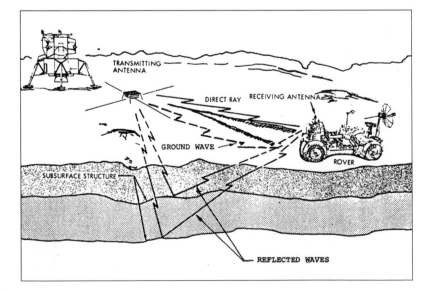

schedule. It had been a tremendous team effort where the medical imperatives had outweighed the urgency for engineering or science and the crews were better for it.

The wake-up call went out from Mission Control at about 204hr 45min to the tune of 'The First Time Ever I Saw Your Face', the start of a busy day in lunar orbit. For five days Ron Evans had kept the science of Apollo alive from above the lunar landscape by methodically working with a separate communications channel, monitored and administered by his own flight director and capsule communicator, operating the SIM bay scanners.

The crew got to sleep at around 218hr for their final night in lunar orbit and were awoken at 225hr 30min with the song 'Come on Baby, Light my Fire', appropriate for the day they would fire up the SPS engine and leave for home. With preparations under way early, the big SPS was fired up at 234hr 02min 09sec for a burn lasting 2min 23sec, executing a ΔV of 3,046ft/sec (928m/sec). The burn was so accurate that the entry gamma was -6.31° and only a small course correction would be required on the way back.

As Mission Control acquired signal around the eastern limb of the Moon 13 minutes after the burn began, the spacecraft was about 335ml (538km) above the crater Tsiolkovsky – around on the far side of the lunar sphere and unseen from Earth. But live television from Apollo 17 gave Earth viewers a chance to see it as the spacecraft climbed out away from the surface.

For some time the TV camera showed the receding Moon, as Apollo 17 climbed up out of its gravity well and a series of tasks occupied the crew before they settled down again for their first rest during trans-Earth coast. That sleep period started at 240hr 30min, the equigravisphere being crossed at 247hr 59min 50sec when the spacecraft was 38,922ml (62,625km) from the Moon and 197,469 ml (317,727km) from Earth and now speeding up ever so slowly at first with less than 54hr to go before splashdown.

As they slept, in Mission Control there was a sense of sadness that each event being ticked off as accomplished was the last time they would be tasked with this particular type

of challenge. But there had been relief that the ascent from the surface had gone off as though it was an everyday occurrence, and that the Trans Earth Injection burn had done its job, the SPS engine now no longer needed, all the worry that it might fail assuaged. Never again would an Apollo spacecraft be this way and never again would Apollo astronauts return from deep space. In fact, it would be more than 50 years before astronauts did so again.

Preserving the tradition, at 247hr 50min the crew were awoken to the rhythm of 'Home for the Holidays', to which the response from Apollo 17 was an appreciative wish for Mission Control to choose that sort of music every morning, attracting a pithy response: 'Well, if I'm waking you up on Wednesday morning, fella you're in trouble!' Splashdown was set for shortly after noon, Houston time, on Tuesday 19 December.

Ron Evans got to be a spaceman when the trans-Earth EVA began at 254hr 54min 40sec, lasted 1hr 5min 44sec and saw the Command Module Pilot retrieve the valuable cassettes containing the data he had been so active in obtaining during much of the period in lunar orbit. During this day the last of six seismic profiling charges set out across Taurus-Littrow valley was detonated on the Moon, providing data that would be used to effectively probe the top 2ml (3km) of the lunar surface at that site.

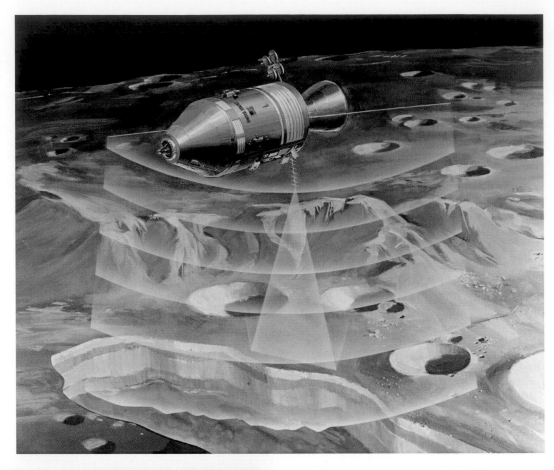

RIGHT The Lunar Sounder was unique to the Apollo 17 mission and helped define the properties of the surface and the interior of the Moon using HF and VHF radio signals. *(NASA)*

BELOW Ron Evans becomes a true spaceman as he retrieves cassettes from the Scientific Instrument Module, and seen here carrying the Metric Camera film back to the Command Module. *(NASA)*

The crew turned in for their penultimate sleep period in space at about 265hr 30min.

To the tune of 'We've Only Just Begun', the wake-up call went to Apollo 17 at 272hr 20min and the crew busied themselves over the next several hours before a televised press conference began at 281hr 27min. Questions submitted by the news media were read to the crew for answers, beginning with Jack Schmitt and his impressions, as a geologist, of the lunar surface, followed by the feelings of the crew regarding their experiences; what was the most memorable event, how did the crew feel about the end of Apollo lunar exploration, and about flagging interest in the United States since very little of the mission was shown on live TV? During the 26min telecast, all three crewmembers proved to be diplomatic respondents, although when asked whether NASA had waited too long to send a geologist to the Moon, Schmitt replied: 'I think the United States waited too long to go into space in the first place and I think we're probably going to wait too long to go back!'

The final sleep period of the mission began at 288hr 30min and lasted a little under seven hours before they were awake and preparing for the last course correction, a 9sec firing of the RCS thrusters at 298hr 38min 01sec for a ΔV of 2.1ft/sec (0.64m/sec) resulting in an entry gamma of -6.73°. The Command Module separated from the Service Module at 301hr 23min 49sec followed by entry interface just under 15min later.

The communications black-out period caused by a plasma sheath around the conical spacecraft began 17sec after that and lasted 3min 20sec, some 4min 5sec before the forward heat shield covering the recovery compartment was jettisoned and the three drogue parachutes were deployed two seconds later. The main parachutes popped out just 4min 46sec before splashdown, which occurred at 301hr 51min 59sec – at 12.5 days, the longest Apollo flight of them all.

The spacecraft had landed in the Pacific Ocean close to the USS *Ticonderoga* and remained apex-up for crew retrieval, with them arriving on the deck of the aircraft carrier 52min after splashdown, their spacecraft

following just over an hour later. With biological isolation having been abandoned with Apollo 15, both formal and informal procedures followed, the crew arriving back in Houston during the morning of 21 December shortly after the first samples.

The third J-series mission had returned a greater abundance of treasure than either of the two preceding missions, the 142 lunar samples collected weighing 243lb (110.4kg) including a wide range of rock types, glass samples, a permanently shadowed soil sample, three double core tube samples, two single core tube samples, a vacuum sample and a drill core sample from a depth of 9.8ft (3m). The SIM bay instruments had been similarly fruitful, the panoramic camera having taken 1,603 frames, the mapping camera more than 3,000 frames and the laser altimeter some 3,769 shots of accurate height data. The Infrared Radiometer had obtained more than 100hr of data, the Lunar Sounder, although plagued by a problem with the antenna extension/retraction mechanism, obtained more than 10hr of data, and the UV Spectrometer worked in lunar orbit for 114hr.

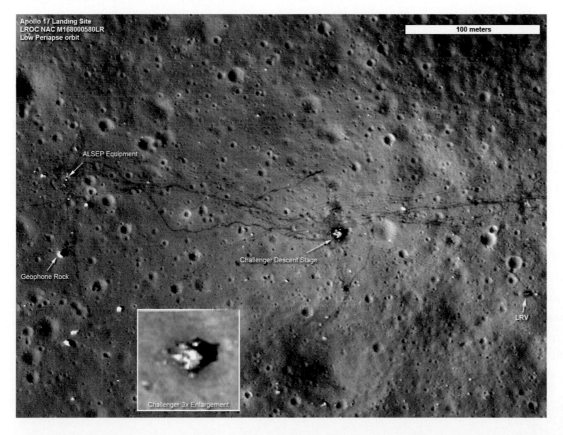

Apollo 17 Landing Site
LROC NAC M168000580LR
Low Periapse orbit

100 meters

ALSEP Equipment

Geophone Rock

Challenger Descent Stage

LRV

Challenger 3x Enlargement

LEFT The Apollo 17 site as seen by NASA's Lunar Science Orbiter several years later, clearly showing the location of hardware and the LRV tracks around the Lunar Module and the ALSEP deployment area. *(NASA)*

5 Going back

The Apollo programme remains the only human endeavour to succeed in sending teams of astronauts to the Moon. Now, more than 50 years later, there are concerted efforts to send people back but this time through an international endeavour which may establish the first permanent scientific bases on another world.

OPPOSITE NASA is developing the Orion spacecraft. Built by Lockheed Martin and the European Space Agency, it is capable of carrying four people for up to four weeks and could be used to support a return to lunar orbit in the 2020s. *(NASA)*

When the crew of Apollo 17 returned to Earth the US space programme had changed for ever and would be re-born from a clean slate, a series of initiatives that would carry the space programme into a new century and a new era.

Gone were the old Saturn rockets, a single two-stage Saturn V remaining to launch the Skylab space station on 14 May 1973 and three smaller Saturn IB rockets to deliver three teams of three astronauts to maintain habitation until early 1974. This would be followed by the last Saturn IB to send the last Apollo spacecraft to dock with a Russian Soyuz spacecraft in June 1975. It was an international venture to consolidate co-existence and détente after a substantial set of arms control agreements signed three years earlier set limits on nuclear arsenals.

In place of Apollo, by the end of 1972 contracts were in place to build a reusable Space Transportation System (STS) as a fundamental shift in direction – to go back to basics and build in low Earth orbit a permanently manned facility where teams of astronauts from several countries could live and work. But there were delays and disappointments. Originally expected to fly in 1977, the Space Shuttle would not see its first flight before 1981. And then, with an agreement reached in 1984 to build a permanent space station embracing partners from Europe, Japan and Canada, a near collapse of that plan threatened its realisation until, after the end of the Cold War, Russia became a partner in the mid-1990s. The first element was launched in 1998.

The last of 135 Shuttle launches flew in 2011, 40 years after the start of the Apollo J-series missions, with completion of what was now called the International Space Station. That assembly of modules and research laboratories has been permanently manned since 2000 and is likely to remain in operation into the 2030s, although it is only formally approved by all the partners to 2024.

BELOW The Orion spacecraft compared to the Apollo spacecraft in size and basic parameters. *(NASA)*

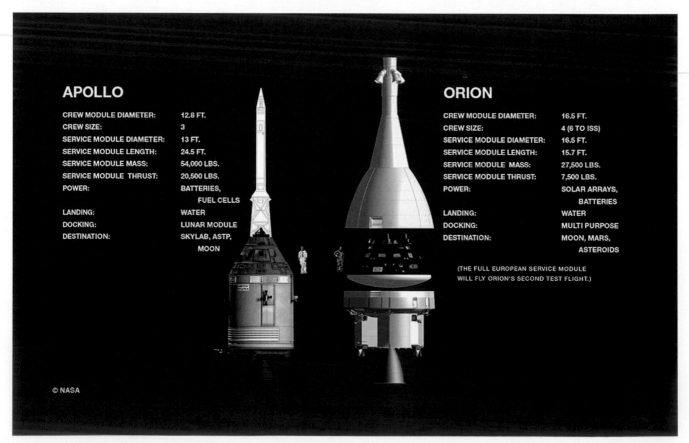

APOLLO

CREW MODULE DIAMETER:	12.8 FT.
CREW SIZE:	3
SERVICE MODULE DIAMETER:	13 FT.
SERVICE MODULE LENGTH:	24.5 FT.
SERVICE MODULE MASS:	54,000 LBS.
SERVICE MODULE THRUST:	20,500 LBS.
POWER:	BATTERIES, FUEL CELLS
LANDING:	WATER
DOCKING:	LUNAR MODULE
DESTINATION:	SKYLAB, ASTP, MOON

ORION

CREW MODULE DIAMETER:	16.5 FT.
CREW SIZE:	4 (6 TO ISS)
SERVICE MODULE DIAMETER:	16.5 FT.
SERVICE MODULE LENGTH:	15.7 FT.
SERVICE MODULE MASS:	27,500 LBS.
SERVICE MODULE THRUST:	7,500 LBS.
POWER:	SOLAR ARRAYS, BATTERIES
LANDING:	WATER
DOCKING:	MULTI PURPOSE
DESTINATION:	MOON, MARS, ASTEROIDS

(THE FULL EUROPEAN SERVICE MODULE WILL FLY ORION'S SECOND TEST FLIGHT.)

© NASA

Rewind

Over the decades, beginning in 1969, high-level plans for a return to the Moon, and a commitment to carry humans to Mars, were laid by one US President after another. None succeeded. Until, on 14 January 2004, President George W. Bush proposed retirement of the Shuttle in 2010 and a programme, called Constellation, to return Americans to the Moon – to stay. On taking office in 2009, President Barack Obama cancelled Constellation, citing cost overruns and unaffordability. But the US Congress pushed the White House to accept development of a deep-space exploration programme based around legacy Shuttle hardware for a launch vehicle known as the Space Launch System (SLS). It was sold on the premise that this would send an enlarged version of Apollo called Orion – the conical command module built by Lockheed Martin and its service module built by the European Space

Agency (ESA) – on missions to the Moon, the asteroids and possibly Mars.

The Orion spacecraft is capable of carrying only four astronauts and supporting them for up to one month and so it is not capable of carrying people to Mars and back. That would require a separate habitation module. But a plan has emerged since 2016 for using the SLS and the Orion spacecraft to support development of a man-tended facility in lunar orbit which would be periodically visited by astronauts supporting a diverse range of crewed and uncrewed vehicles for landings on the Moon. Many of these vehicles would be provided by commercial companies leasing their vehicles to NASA.

ABOVE Called the Lunar Gateway, Orion and the partners on the International Space Station (NASA, the European Space Agency, Russia, Japan and Canada) would assemble and operate a staging post in lunar orbit from where manned and unmanned landers would support a permanently manned Moon base. *(NASA)*

On 11 December 2017, President Trump signed his Space Directive No 1, which required NASA to return to the Moon and do that by exploiting commercial carriers – both launch vehicles and spacecraft – as had been successfully achieved with the International Space Station. The same policy plan initiated by President George W. Bush had started a commercial programme for carrying cargo to and from the ISS and two companies successfully competed to deliver that service. This was followed by a further programme to carry people to the ISS and two companies, SpaceX and Boeing, are developing spacecraft to achieve that. Until they do, all personnel

flying to and from the ISS do so using Russia's Soyuz spacecraft.

With the commercial launch vehicle and spacecraft providers now underpinning logistical and human support for the ISS, the US government plans to expand that model and use it as a template for commissioning further development to support the lunar-orbiting facility – called the Gateway – and the vehicles to go down to the surface and operate robotically, or with crews, to begin a permanent presence on the Moon in much the same way the international partnership has achieved for the ISS.

Renewal

In March 2019, NASA pledged to return humans to the Moon as a priority using lunar bases for the scientific exploration of our nearest neighbour and to begin the technological development of a resource asset to provide materials that could sustain the manufacture of facilities and a support infrastructure using 3D printing. This would avoid the need to lift large quantities of materials from Earth, totally re-imagining the future Moon bases that 50 years ago would have required large modules, habitats, laboratories and power supplies lifted from Earth. Now, all that may be required are the 3D printers utilising in situ regolith, Moon rock and minerals together with possible water resources which are believed to exist near the poles and in permanently shaded places.

The future programme, of a lunar Gateway in orbit and the initial settlements on the surface, is a goal jointly shared by the ISS partners, who have all pledged to support this work, which has bi-partisan support in Congress and a firm direction from the White House. The consistency that now underpins a renewed approach is essential for returning humans to the Moon. While it may depend to a large extent on the commercial providers to ensure the infrastructure is in place, these providers are 'hired contractors' and without strong agency and national political leadership to provide direction and the money to achieve the ambition, it would not happen.

In the future, resource mining of lunar assets may be self-sustaining but for the present that is a hope. At last, however, a firm national policy direction well mapped, effectively laid out and properly funded has been agreed. The money which NASA receives is but 0.4% of US federal government spending each year and it is unlikely that this will increase. But, with NASA's space programme returning ten-fold to the national economy, it is a programme which is prudent, effective and well managed, providing stimulus to a base of commercial providers who would not be there were it not for these goals – goals which are now truly international.

The first and the last men to walk on the lunar surface are no longer with us, but the motivations that pushed them and the others who flew to the Moon, to meet national ambitions and challenge their capabilities, are still there in a partnership between government and industry, alive with new talent and a solid determination to pick up where Apollo left off.

BELOW Planned for initial assembly beginning in 2024, shown here are the separate elements of the Lunar Gateway, which would be supported logistically by commercial providers. *(NASA)*

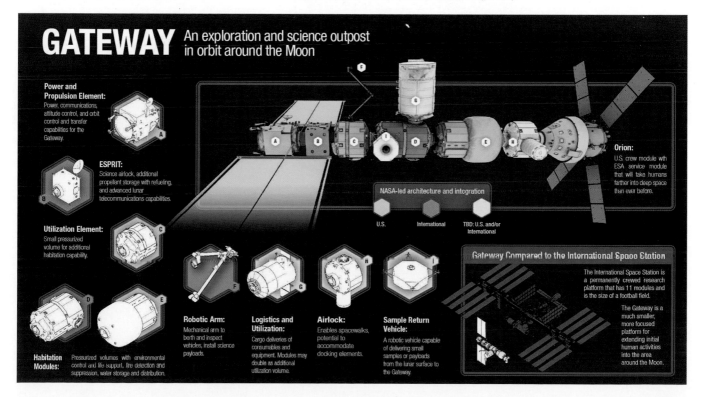

The last man on the Moon during the Apollo Age of Exploration, Eugene Cernan with the third of three Lunar Roving Vehicles. *(NASA)*